Muslim Festivals in India
and Other Essays

Garcin de Tassy

Translated and edited by

M. Waseem

DELHI
OXFORD UNIVERSITY PRESS
BOMBAY CALCUTTA MADRAS
1995

Oxford University Press, Walton Street, Oxford OX2 6DP

Oxford New York
Athens Auckland Bangkok Bombay
Calcutta Cape Town Dar es Salaam Delhi
Florence Hong Kong Istanbul Karachi
Kuala Lumpur Madras Madrid Melbourne
Mexico City Nairobi Paris Singapore
Taipei Tokyo Toronto
and associates in
Berlin Ibadan

ISBN 0 19 563677 5

Laserset by S.J.I. Services, B-17 Lajpat Nagar Part 2, New Delhi 110 024
Printed at Wadhwa International, New Delhi 110 020
and published by Neil O'Brien, Oxford University Press,
YMCA Library Building, Jai Singh Road, New Delhi 110 001

Muslim Festivals in India

Preface

What follows is a translation of Garcin de Tassy's *Mémoire sur des particularités de la religion musulmane dans l'Inde, d'aprés les ouvrages hindoustani* (Paris, 1831) coupled with his reviews of Mrs Hassan Ali's *Observations on the Musulmauns of India, etc.*, (1932) and Ja'far Sharif's *Qanoon-i Islam*, etc. (1832)—reviews which appeared respectively in *Nouveau journal asiatique* (June, 1832) and the *Journal des savants*, of August and September, 1833. He also wrote a *Notice sur les fêtes des Hindous, d'après les ouvrages hindoustani* (*Nouveau journal asiatique*, February and March 1834), where he suggested that the four pieces should be treated as a whole. However, I have dropped the *Notice*, which does not fit the pattern I had in mind. Very sketchy, repetitive if not monotonous, it is a catalogue of Hindu festivals classified according to the lunar months and is based largely on the first volume of *Hindee and Hindoostani Selections* by Tarini Charan Mitr and W. Price of Calcutta. Significantly, the most interesting parts in the *Notice* are quotations from the Urdu books de Tassy has used in the *Memoir*.

For those interested in the contemporary Indian background' and an explanation of the names and terms in de Tassy's text, I have added an Introduction and an Appendix on Indian saints. De Tassy's spelling of non-European words has been modernized but the original spelling may be found in the Index. I have converted into Roman script the words and sentences de Tassy gave in the Perso-Arabic script. Similarly, in the interest of readability, I have modified de Tassy's use of italics, and the repetition, in the Perso-Arabic script, of certain words which had already been given by him in transliterated from.

Thanks are due to the British Library, Print and Drawing Section for the permission to reproduce the painting entitled 'An offering to the Ganges'.

I am indebted to Professor Nazir Ahmad of Aligarh, Professor Shoeb Azmi, Professor Mujeeb Rizvi, and Mr Shahabuddin Ansari of the Jamia Millia, whom I consulted on several occasions for their help; while Ahmed Ali's excellent translation, *Al Qur'an* (Oxford University Press, 1987) is the source for the phrases from the Holy Book.

Thanks are due to Dr A. R. Bedar of the Khuda Bukhsh Khan Library, Patna, for providing me with a copy of the *Memoir* and other related material, and to Anne Renard, daughter of my friends in Versailles, for getting me promptly a copy of de Tassy's review of Ja'far's book from the Bibliothèque Nationale.

I am grateful to Professor Sunanda Datta of Calcutta for the information she gave me on the dress of the women in the painting, 'An Offering to the Ganges'. I must also record my gratitude to the Trustees of the Charles Wallace Foundation, London, and La Maison des Sciences de l'Homme, Paris, for their hospitality for a different project; a visit to these cities helped me in making slight but very significant changes in the Introduction and the Appendix. I am also grateful to Mr Vasant Sathe, Chairman, Indian Council of Cultural Relations and Professor Bashiruddin Ahmed, Vice-Chancellor, Jamia Millia Islamia for the travel grant.

I am really grateful to Dr J. M. Lafont, who not only provided me with the necessary material but also initiated me into the art of translating scholarly works; to Dr Narayani Gupta for suggesting that this translation was worth publishing, and to my publishers for agreeing with her.

Incalculable and inexpressible, however, is my debt to my daughters, Saba and Saleha, and their mother, Adiba, to whom I dedicate my contribution to this book.

Jamia Millia Islamia   M. W.
New Delhi

Contents

MEMOIRE

SUR DES PARTICULARITÉS

DE LA RELIGION MUSULMANE

DANS L'INDE,

D'APRÈS LES OUVRAGES HINDOUSTANI,

PAR M. GARCIN DE TASSY.

PROFESSEUR D'HINDOUSTANI À L'ÉCOLE ROYALE ET SPÉCIALE DES LANGUES
ORIENTALES VIVANTES, MEMBRE DES SOCIÉTÉS ASIATIQUES DE PARIS,
DE LONDRES ET DE CALCUTTA.

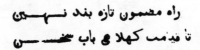

راه مضمون تازه بند نهین
تا قیامت کهلا هی باب سخن

On peut dire encore quelque chose de
nouveau; que dis-je? la porte du discours
restera ouverte jusqu'au jour de la résur-
rection.

WALI.

PARIS,

DE L'IMPRIMERIE ROYALE.

M. DCCC. XXXI.

Introduction

GARCIN DE TASSY's *Memoir* is an early but pioneering work on syncretism in Indian Islam, a subject which is rather popular with western researchers. If one goes by the number of editions published, it has been the best received of de Tassy's books, with his translation of Fariduddin Attar's *Mantaqut tair*.[1] A sequel to an earlier piece on the doctrines and duties of Muslims, according to the Quran, *Doctrine et devoirs de la religion musulmane, tirés textuellement du Coran, suivis de l'eucologe musulman traduit de l'arabe*, published in Paris, 1826 the *Memoir* is an attempt to show how Muslims in India have adopted certain religious practices not found in other Muslim countries, which have no sanction in the Quran, and which are often in contravention of the Muslim scripture. At the same time, de Tassy traces certain words and ceremonies to non-Muslim western sources; for example, the origin of the Hindustani word, *nau* (boat) and the custom of setting afloat in rivers small boats with lamps in memory of Khwaja Khizr.

However, we now know that what was Hindu could be Muslim, what was Muslim could be Hindu, or that it could be both or neither. For instance, the spiritual exercise of breath control (*hubs-dam*) one of the most important features of Indian sufism could be either imported or indigenous; so could be *chilla-m'kusa*, hanging upside down in a dark place for forty days in order to meditate. Whatever its origin, it was practised by Fariduddin Ganjshakar, as well as by devotees in Bahra'ich and as Schimmel reports, 'it is still today sometimes performed by dervishes who may use the hat rack of a Pakistani train to hang from' (p. 132). Millions of Hindus visit Bahra'ich, the site of the oldest Muslim shrine in India, to celebrate the preparation of the the marriage of Salar Mas'ud, who died fighting the Kafirs. As de Tassy has shown, a Hindu Teli (oil man) brought a father-in-law's marriage gift for the hero, who was slain on the day of his wedding with a Muslim lady Zuhra Bibi, who is supposed to be buried by his side (Schimmel, p. 135). Irreverent practices, exhibitionism, curious and funny feats, crowds of women, and the use of intoxicating drinks were common. It has also been suggested that the shrine was originally a sun temple. However it may be, the festivals described by de Tassy have now assumed greater social and political importance and much larger mass appeal. New *pirs* rise and fall and old ceremonies have acquired new significance. The twelfth of Rabi'ul Auwwal was observed in India as *Bara wafat* (literally, death on the twelfth) as a death anniversary (greatly perplexing de Tassy) with very solemn and private prayers.[2] Now the Birth of the Prophet (Id Milad-un Nabi) is celebrated as a public festival, and the Central Government has grudgingly included it in the calendar of public holidays. Streets are now lighted and decorated and processions taken out—one of these processions led (in 1992) to riots in the otherwise communally peaceful city of Madras. The veneration for the Prophet's footprint could be traced equally to a Meccan tradition as well as to the influence of the worship of the footprint of Buddha and Vishnu. However, a reader's assertion in a letter to the Editor (*The Times of India*, 13 October 1992) that there are three footprints of Vishnu, one each in Bodh Gaya and Shukla Teerth, and the third one in the Ka'ba, would not have surprised many readers.

The secular King of Oudh, Asafuddaulah, would never miss the annual celebrations at Bahra'ich, and the equally secular Prime Minister of India, Pandit Nehru, would grant *darshan* to millions of pilgrims in Allahabad during the Kumbha mela, leading, incidentally, once to a stampede, which left hundreds dead. When Haji Ilyas, King of Bengal, marched to Bahra'ich under the pretence of seeking a cure for leprosy, there was panic in Firoz Shah Tughlaq's Delhi; it was feared that the Bengali might attack Delhi on the plea of offering prayers at the tomb of Nizamuddin Awliya. One can imagine the apprehensions of the leaders of the National Front when the Congress Party decided to hold its first post-election meeting (in 1992) in the temple city of Tirupati. *Plus ça change plus c'est la même chose.*

De Tassy's basic argument and most of his facts are irrefutable, but a few errors need to be corrected, some gaps to be filled and many entries re-arranged in order to do justice to the pioneering work, which has been mentioned in every work on popular Islam in India.

In support of his argument de Tassy depended almost entirely on whatever the Indian Muslims themselves had written on their festivals. Since most of these books were commissioned by Fort William College, which was controlled by alien masters, the writers were, however, either unable or unwilling to tell the whole truth. Secondly, de Tassy did not have sufficent contact with Indians while writing this *Memoir*, to be able to check his sources. Hence, there are unfortunate errors, lack of understanding of various religious Orders, of the kinds of saints and of the class structure of Indian Muslims.

Since this topic is both a vast one and of great popular interest I shall confine myself to the books by Ja'far Sharif[3] and Mrs Hassan Ali, which were in press when the *Memoir* came out. De Tassy missed the very interesting account in Persian written around 1812 by Muhammed Hasan Qateel (né Dewani Singh, 1758–1817) on the ceremonies and the social and caste structure of both Hindus and Muslims. I will quote liberally from the Urdu translation of this book, *Haft Tamasha*, by Muhammed Umar (Delhi, 1968). For scholarly support to my observations, I have confined myself, largely, to *Islam in the Indian Subcontinent* (Leiden, 1980) by Annemarie Schimmel, who in her scholarship, range of interests and sympathies is very close to de Tassy. My other authority is *Indian Muslims* (Delhi, 1985) by Professor M. Mujeeb.

Not much is known about the personal life of de Tassy (1794–1878), who was the son of a merchant in Marseilles—a Mediterranean port with ancient contacts with Arabs. In 1817, he came to Paris, where the well-known Orientalist, Silvestre de Sacy (1763–1824), encouraged him to learn the 'Islamic languages', Arabic, Persian, Turkish and Urdu.'Indianism', specialized interest in Indian languages and literatures, had developed in France in the eighteenth century, one of its pioneers being Anquetil Duperron (1731–1805), who was in India for five years and introduced the Upanishads to Europe with his Latin version of the Sanskrit classic. Like many other French Orientalists, de Tassy was self-taught, an *autodidacte*; he learnt Urdu through Gilchrist's Urdu grammar. (De Sacy had also written a book on Urdu grammar.) Through the good offices of de Sacy a Chair in Urdu was established in the Ecole des Langues Orientales Vivantes in 1828, a post which de Tassy occupied till his death. The chair was then abolished for want of students, many of whom were English boys, who were interested in the language in order to seek a fortune in India. Louis Mathieu Langlès (1763–1824), whose edition of Chardin de Tassy quotes quite often in the *Memoir*, became the founder-director of the Ecole (established 1795) where de Tassy taught. Jean Chardin was incidentally, a traveller, who had visited India and Persia. Since Langlès was a Protestant, a fact repeatedly mentioned by de Tassy, he feared persecution in France, and so spent the remaining part of his life in England where he worked for the Dutch East India Company.

Most of de Tassy's writings were published in journals, very little was reprinted and even less translated, that is why his works are not easily available. However, a look at the bibliography of his writings would show that his main interest was Islam and Islamic literatures, and naturally enough there is a constant comparison in his writings with Roman Catholicism and Western literatures. From a certain point of view de Tassy's most important contribution to scholarship in this area was the Annual Inaugural lectures which he delivered in the first week of each December from 1850 to 1877. They record not only his intellectual growth and the focusing of his interests, but also take notice of all important books published in India and the changing climate in which Urdu was developing, particularly in the light of the growing antagonism from supporters of Hindi.

The *Memoir* is a very early piece. Apart from translations from the Arabic and Turkish, de Tassy had written on doctrines and duties in Islam,[4] and had translated some extracts from Afsos' *Araish-i Mahfil, The Birds and Flowers* by Izzuddin Moqaddasi, as well as Mir's satire on bad poets. He was interested in Islam as it was actually practised in various countries, and in how it compared with the teachings of the Quran. Secondly, he insisted that the fulness of a culture could be understood not through an Orientalist's or official's or traveller's observations, but only through the nuances coded in the idiom of that language.

Unfortunately, there was very limited material for such an approach; Fort Willliam College had just been established, but for the College authorities Islamic literature was not of prime interest. De Tassy could not visit India, perhaps on account of lack of financial resources, but he succeeded remarkably because of his imaginative, yet honest and very practical approach to the subject.

Like Schimmel (pp. 120–3), de Tassy found it convenient to follow the *Bara Masa* tradition of classifying material which in his case is mostly translations from Urdu books. However, since not all Muslim festivals are celebrated according to, what is in popular parlance called 'Urdu months', but sometimes according to 'Hindi months', de Tassy had a problem. Jawan had evolved a compromise by choosing a particular solar year and describing all the Hindu and Muslim festivals which were held in that year. De Tassy was not satisfied with this method and that is why he divided his material into three parts. However, he made a serious mistake by introducing Part II, where he described saints for whom he thought there were no annual celebrations.[5] Every saint belonging to a mystical order must have a *mazār*, and wherever there is a *mazār* there must be an *urs*.

De Tassy is mistaken in saying that the Muslim calendar is 'lunar'and the Hindu 'solar'.[6] As a matter of fact, the 'Hindu months' are also 'lunar', with the difference that the month begins at an astronomically calculated point of time when the crescent moon appears, whatever the time of day or night. On the other hand, the 'Urdu months' begin when the moon is sighted on the western horizon after sunset; if it is not visible during that short period of time, the month begins the following evening. Secondly, in order to adjust it with the solar year, on which the cycle of seasons so loved by Hindus, depends, one extra month, *malmas*, is added after every third year.

Before we proceed, a few words on the translation of *'fête'* and *'musulman'* in the title of the *Memoir*. Except for Moharram, which is not celebrated but 'observed', as a day or period of mourning, 'festival' can be an appropriate translation of 'fête', which is celebrated by a community as a special day with varying degrees of joy and solemnity. However, I have retained de Tassy's 'fête', often arbitrarily, when it is a question of private, domestic religious ceremonies; ceremony, in the sense of what has come down, but is not against the words of the Quran and the Hadith; it has nothing to do with 'worship' and 'ritual.' We do not worship God, we pray to God, similarly there is something pagan about 'ritual' if it means repetition of certain words and movements the significance of which we do not know. But everything—making amulets with the words of God or asking a devotee to wash holy writings in water and drink it—is permissible if it creates faith or at any rate if does not endanger it.

It was felt that '*musulman*' should be translated as Muslim, since the word has a history in India, at any rate. The French have, since de Tassy, retained the spelling, the pronunciation and perhaps the connotation of the word, but in English there are several spellings of the word. Moreover, in the later half of the nineteenth century 'Mohammedan' was revived to refer particularly to the educated and 'respectable' class of the community, but this English construction was not acceptable to the Faithful, since they believed—a fact repeated by de Tassy—that Muhammad did not start a new religion as Christ did. Thus, the Mohammedan Anglo-Oriental College, Aligarh (established in 1887) became Aligarh Muslim University in little more than a decade after the establishment of All India Muslim League in 1906. The Arabic 'Muslim' superseded the Persian 'Musulman', although the followers of the Prophet continued to call themselves Musulman in Urdu. But in English, as politicians were aware, Musulman retained an emotive connotation suggesting the national or international community (*umma*).

I have, however, retained de Tassy's *Hindustani*, a term which expresses a particular concept of the language and has now a political history, from Gandhi's experiment with a language written in the two scripts, Persian and Devanagri, to Nehru's wish to have a language without the arbitrary insertion of unfamiliar Sanskrit words. De Tassy was aware of the word Urdu, when writing the *Memoir*, for the full title of Mir's collected poetry was *The poems of Meer Mohummud Tuqee, comprising the whole of his numerous and celebrated compositions in the Oordoo or polished language of Hindoostani Ed. by learned Moonshees*, Calcutta, 1811.

After the decline of Persian and before the rise of Hindi and English, Urdu (de Tassy's *Hindustani*), was the literary language of the élite, wherever Muslim influence reached, in the Indo-Gangetic plains, in the southern states ruled by Muslims, and in certain other pockets, where migrants from these places had settled. As the etymology of the Turkish word (*Urdu*, a Tartar Khan's camp, from which we have the English 'horde'!) would suggest, Urdu began as a bazar language of Delhi, a product of practical everyday needs, a language where the structure of the sentence remained native and foreign words, mainly Persian, were introduced according to necessity. However, in the North, the language of discursive prose continued to be Persian till the first few decades of the nineteenth century. But thanks to the efforts of *sufi* saints, Urdu attained the status of a vehicle for literary expression quite early in the Deccan, with Gesudaraz Bandanawaz (1321–1422) supposed to be the first prose writer in Urdu. Interestingly, de Tassy's favourite poet, Wali Dakani, and first poet of the North Indian style, *raikhta*, was born in Aurangabad in the Deccan, and it is doubtful if he ever visited Delhi. However, his verses created a sensation when they reached Delhi; during Holi, Kayasthas would sing verses from the *Gulistan* and the *raikhta* of Wali in the streets (Qateel, p. 93.) Wali's contribution was twofold: he minimized the linguistic differences between northern and southern styles and transferred for the first time to Urdu, the prevalent sensibility which was predominantly Persian and which had found expression in Indian poetry written in Persian.

To Wali's age belonged Fa'iz, whose given name may not have been Sadruddin Muhammad, as de Tassy would suggest.[7] The first poet to write in Urdu from North India, he belonged to the aristocracy and got his rather slim volume of verses compiled in 1714. He is distinguished from what later became the mainstream of Urdu poetry, not only in his choice of subject, but also in his use of local, non-literary images, words and rhythm. A dominant theme in his poetry is the sensuous beauty of women who appeared in public. He has written about girls who assemble at wells to draw water, who go to the Nagambodh ghat for a holy dip, who sell fruits in the lanes, and and at stalls sell *biras* of pan.[8] For his short poems as well as *ghazals*, he chose the convention of *sarapa*, in which the jewellery and costume worn by the girl on different parts of her body are described. 'In Praise of the Woman Bhang-seller of Dargah Qutub', from which de Tassy has quoted above, is mainly a description of the beauty of this 'low-born' woman. The narrative, the atmosphere and the moral at the end of the poem are invented,largely, to provide a background to the portrait of the the beautiful girl.

It appears that with the work of Mir Taqi Mir (1724–1810), Mirza Rafi'uddin Sauda (1713–81) and Mir Hasan (1727–86), the line along which Urdu poetry was to grow had been determined; they have become part of the tradition, in the sense Fa'iz has not. It was during their time that the collapse of Delhi as a cultural centre became complete and poets began their journey eastward, creating local centres all the way from Lucknow to Banaras, Faizabad, Patna, Murshidabad and finally at Calcutta.

De Tassy is full of admiration for the Urdu poets he knew; he was not aware of Ghalib, then. He uses a verse from Wali for the epigraph and rounds off the *Memoir* with a verse from Mir Taqi, but these poets were hardly of any help to him in his research. In the first place, poetry is generally not the source for the kind of material de Tassy was looking for. Secondly, the urban-based Urdu *ghazal* poet of the North was expected to draw upon the repertoire of imagery supplied by Persian poets. Non-*ghazal* verse was too serious in its themes to cover details of the daily, mundane trivialities. Fa'iz alone was an exception. A word on de Tassy's comparison of the ghazal with the sonnet (see p. 44 below).[9] In most of the *ghazals*, and in the case of Mir almost always, each couplet is rather like a two-line poem, which is complete in itself; what holds the disparate couplets together are the meter and rhyme. Fa'iz again is an exception; he describes feelings about women in logically connected sets of couplets.

Since the poets he so greatly admired were of so little help to him, de Tassy had to depend heavily on quasi-literary sources for his *Memoir*.

Of the ten books mentioned by de Tassy as his sources, eight were published in Calcutta under the guidance of John Gilchrist (1759–1841), Professor of Hindustani in Fort William College, established in 1800 and closed in 1854. Gilchrist's association with the College was brief; he went back to England in

1804. But he had gathered around him talented writers who through their original writing and translations would give to the Company 'writers' (young apprentices), the English specialists and bureaucrats, a picture of India as it appeared to the Indians themselves. The irrepressible use of figures of speech and reproduction of unchecked facts were condoned because these books were considered essentially raw materials with which the British would form their own opinion about India. An interesting exercise would be to compare *Araish-i Mahfil* with Hamilton's *Gazetteer*, where notes on various places are given in alphabetical order and at the end of each note the source is indicated. Afsos does not do any such thing.

However, Gilchrist is often unfavourably compared with de Tassy on account of having been guided by imperialist interests rather than by a single-minded love for India or Urdu. Moreover, he is blamed for sowing the seeds of the Hindi-Urdu controversy, which is often considered to be a factor leading ultimately to the division of the country.[10] On the other hand, de Tassy is regarded as a sincere and consistent benefactor (*mohsin*) of Urdu; he had no political axe to grind; he started writing long after the French had ceased to have imperialist interest in India. Ironically, however, de Tassy had to depend upon material supplied by the College, established on the first anniversary of the Battle of Serangapatnam, which was believed to have sealed the fate of the French in India. Secondly, his scholarship was used mainly by the British.[11]

In justice to Gilchrist, he was a man of literary taste with definite views on the relationship between culture and language, though he wrote only a few elementary books on the Urdu language. Moreover, for the first time in the history of Urdu, he inspired and supervised the compilation, composition and translation of a very substantial body of literature, which researchers like de Tassy could make use of.

All the four writers who provide the bulk of source material for the *Memoir* were full-time munshis (translators, compilers and teachers) of the College ; all of them were born in Delhi, but died in the East. Haidar Baksh Haidari (1768–1828) came to Calcutta via Banaras, where he returned to die after retirement. *Gul-i Maghfirat* is the summary of his own translation of a Persian work, *Rauzat-ush Shuhuda*. Commissioned by the College, it contains sixteen *majlises* instead of the usual ten. *Majlis* (literally, assembly) in this context means a lecture dealing mainly with the sufferings of the House of Husain. Quotations from Jawan (pp. 50–1 below) can give us an idea of the theme and atmosphere of such assemblies.

The author of *Bara Masa*, Mirza Kazim Ali Jawan (d.1816) came to Calcutta via Lucknow and Banaras. Commissioned by Gilchrist, the *Bara Masa* gives a chronological description of various Hindu and Muslim festivals.[12]

Amanatullah Shaida came direct to Calcutta, where he died in 1261 H (1845–6). *Hidayatul Islam* is a collection of Arabic prayers (with interlineal meaning in Urdu), which Muslims recite on different occasions, in addition to the formal five obligatory prayers called *namāz*, or *salāt*. In the Preface, Shaida explains the reasons for the Urdu translation giving the example of the shift from Latin to the national languages of the European Middle Ages. (However, these prayers must be recited in Arabic, the Urdu translation is given to make the faithful understand what he was really reciting—a compromise between the modern Christian and the traditional Hindu practice.) This book was translated by de Tassy, who was greatly fascinated by it. Gilchrist who had great respect for Shaida's scholarship asked him to translate the Quran, part of which was also published but was finally suppressed by the authorities in 1804.

Mir Shair Ali Afsos (1747–1808), de Tassy's fourth source, wandered in different parts of eastern India, Allahabad, Patna and Banaras, till in Lucknow he got an appointment to the College in Calcutta, where he stayed till his death. One of the most popular publications of the College is his *Araish-i Mahfil, being a history in the Hindoostani language of the Hindoo princes of Dihlee from Joodhishtur to Pithoura, compiled from the Khoola-sut-ool Hind and other authorities*, Calcutta, 1808. In his work Afsos updates Subhan Rai Bhandari Batalvi's Persian text *Khula-sa-tut Twarikh* (and not, as de Tassy records, *Khoola-sut-ool Hind*), which deals with events only up to H. 1107. It was re-edited in 1863 by W. Nassan and translated into English in 1871. In India alone, at least four editions of this work were published in the nineteenth century. The book is a free translation with additional material from the *Ain-i Akbari* and other sources as well as Afsos' own information about persons and places.

Thomas Roebuck (1784–1816) had met Gilchrist in England, when he became interested in Urdu and was with the College from 1810 till his premature death, six years later. One of the most eminent professors of the College, he has written on grammar and lexicography as well as the Annals of the College. His collection of

proverbs from Urdu was published posthumously by
H. H. Wilson.

The 'Low-Born' and the Respectable

A brief description of Indian Muslims, their social
classification and their doctrinal differences strictly in
the context of the *Memoir* could provide a valuable
background to the study of the work. The division
into four 'classes' (de Tassy, p. 14) 'categories' (Schimmel, p. 111), and 'groups' (Herklots, p. 91) is rather
misleading and definitely not comprehensive. However, Qateel is correct in calling them 'four *firqa* of the
ashraf' (p. 129). The crowd of devotees, hardly distin-
guishable from their Hindu brethren, may not have
belonged to any of the four 'upper' castes—Syed,
Shaikh, Pathan, and Mughal—but they were nonetheless Muslims. In Mir's 'Warning to the Illiterate ',
which de Tassy has translated as 'Advice to Bad
Poets', *Conseils aux mauvais poètes* (1825), Mir recalls
the days when low-born, *ajlaf*, were not allowed to
meddle with poetry and masters instructed only the
ashraf (the respectable). A little further on in the poem,
the poet wonders what *ajlaf* have to do with literary
discrimination, what understanding drapers and carders have of poetry. Mir says merely that people
engaged in the lowly professions are *ajlaf*; he does not
say that the four castes mentioned above are *ashraf*.
Qateel offers a more appropriate dichotomy: Old or
Original Muslims (the conquerors and other
migrants) and the New (*nau*) Muslims, recent converts. But not all immigrants had 'high literary
discrimination' nor were all converts carders and
drapers. According to Qateel, the Kumbos were Indian converts who were highly educated and proud

of their caste. Qateel himself was born Khatri but became Mirza Muhammad Hasan, even when he himself knew that Mirza was a title reserved for Mughals. Mir Taqi Mir would claim — rather too often —that he was a Syed but his contemporary Sauda was not convinced and in one of his satires insinuated that Mir was a baker, that is from the family or caste of bakers. A critic who has taken a charitable view of the whole issue would like to set the matter at rest by saying that it was possible that some one in Mir's family had taken up this professon in the past. Qateel has given an interesting account of the difference between caste and class.

Among the Hindus even those who own elephants would not hesitate to get their daughters married to brokers, gram-sellers, and confectioners. In Shahjahanabad there was a Khatri who had elphants and *palkis,* but his brother-in-law's son carried his *attar*-jug during his rides. But among Muslims, persons engaged in low professons—domestic servants, water-carriers, elephant drivers, medicine and perfume sellers, confectioners and bakers are considered *paji* (*ajlaf*).Let alone rich people, even *sharif* persons who draw a salary of five rupees would not consider it permissible to have matrimonial relations in the family of an elephant driver who might be earning even five hundred rupees as salary; who would then consider water-carriers and others engaged in business in the marketplace? As a matter of fact, many rich persons would not allow reciters of *marsiyas* to sit in their presence except during Moharram.... If a Syed adopts the profession of a tailor or a Moghal becomes a baker, a vegetable seller, or a water-carrier, he is not allowed to sit in the presence of his relations or other *sharif* people. There are two kinds of *sharafat*, *nasabi* (according to birth) and *hasabi* (according to status). The Hindus consider the former more important, but among Muslims it is the opposite... It is not possible for a Syed to give his daughter in marriage to another Syed, who runs a medicine shop (pp. 96–7).

However, if both *hasb* and *nasb* had to be combined in the choice of a bridegroom, it created problems particularly for Syed girls, many of whom remained spinsters, a point noted both by Mrs Hassan Ali (vol. I, p. 8) and Schimmel (p. 115.) Thus it was not uncommon to have fake titles of caste being accepted, if this could be hidden from the community . Hence the joke, 'Last year I was a Julaha, a weaver, this year I am a Shaikh and next year, if prices rise, I shall be a Syed' (Herklots, p. 115). Most of our writers, however, only refer to the urban low castes, who were in the service of the aristocracy, and they do not mention the rural low-born, who formed the bulk of the pilgrims to shrines.

While it is not suggested that all those engaged in lowly professions were poor *ajlaf*, there is reason to believe that in the countryside Muslim converts continued to pursue their hereditary professions. The Muslim barber, for example, discharged the same duties among the Muslims as his Hindu cousin did in his community. He belonged to the same rural low class whose women thronged the shrines. Upper class Muslims in the villages would probably not have allowed their womenfolk to join the *melas* and be jostled in the claustrophobic shrine of Salar Mas'ud. However, it is difficult to hazard a guess about the caste of the assailants of the Mughal chronicler, Badauni, who was discovered with a girl in the shrine of Shah Madar (see below, p. 178).

Shi'a vs. Sunni

De Tassy's curiosity about whether a person was Shi'a or Sunni appears rather excessive today; equally, his comparison of the two sects with Protestantism and Catholicism, is inexact. Sunnis cannot be compared with Protestants merely because a section of them had been trying to 'purify' Islam of extraneous customs, when de Tassy wrote the *Memoir*, nor could the Shi'as be considered similar to the Roman Catholics, merely on account of the importance they attached to ceremonies during Moharram. It is rather the *khanqah*, which has some kind of resemblance with the medieval Church.

Mrs Hassan Ali conversed with 'many sensible men of the Mussulmaun persuasion' on Moharram and she was told that 'the pompous display is grown into habit, by a long residence amongst people who make a merit of showy parade at all the festivals' (vol. I, pp. 53–4). She was told that as in the case of ceremonies during marriages which have become almost identical with those of the Hindus, 'It is in habit only, where "faith" is not liable to innovations,— between themselves and the Hindoo population' (vol. 1, p. 55). She says further:

> The most religious men of that faith feel equal, perhaps greater sympathy for the sufferings of the Emaums, than those who are less acquainted with

the precepts of the Khoraun, they commemorate the Mohurrum without parade or ostentatious display, and apparently wear mourning on their hearts, with their garb, the full term of forty days— the common period of mourning for a beloved object—but these persons never join in Mortem, beating breasts, or other outward show of sadness, although they are present when it is exercised, but their quiet grief is evidently more sincere (vol. I, p. 53).

In order to show how similar were the attitudes of educated people of both the sects towards the tragedy of Karbala, I would quote the eminent Orientalist, on what his father Khuda Bukhsh Khan (1842–1908), himself a pious Sunni, thought about Moharram:

On the 10th Mohurrum my father never allowed us children to see the Mohurrum procession, which he regarded as a mockery and travesty of religion, and for which he never found language sufficiently condemnatory. He thought it wicked to a degree to convert the anniversary of one of the greatest tragedies in the history of Islam into a day of carnival and festivity, instead of observing it scrupulously as one of veritable mourning; and perhaps it would surprise the reader to know that up to now I have not seen the Patna Mohurrum procession which, l am told, is almost unique in grandeur and magnificence. On the anniversary of that terrible day he occupied himself in the study of the Qu'ran and other religious books.[13]

Differences between the Shi'as and Sunnis could be doctrinal—and there are differences among several sects of the Sunnis themselves—but they have often led to conflict and violence in the Imperial court as well as in the lanes of Lucknow. De Tassy has mentioned Nurullah Shushtari (1542–1610), who was one of the leading thinkers of Shi'a Islam, and as the Qazi of Lahore was absolutely impartial. However, he became unpopular on account of his strictness and at the instigation of the *ulema*, Jehangir had him flogged to death (Schimmel, pp. 88–9). Moharram riots mentioned by de Tassy were not religious in character—no riots are—and had nothing to do with the feeling of grief. As Mrs Hassan Ali rightly points out, private quarrels are often reserved for decision on the field of Kraabaallah' (vol. I, p. 93), and 'perhaps, the violence of party spirit may have acted as an inducement to the Shiehs, for the zealous annual observance of this period,' (vol. 1, p. I25). Qateel gives another side of the picture: according to him, *ta'zia-dari*—the installation of the *ta'zia* and observance of the customs and ceremonies related to it—was performed by the Shi'as, Hindus and 'low-born' Muslims. Sunni Ashraf would attend *majlises*, recite *marsias* and also weep on this occasion, but they would not instal *ta'zias* at home. However, some uneducated boys from Sunni Ashraf families would in the arrogance of youth and the love for violence, put up *ta'zias* in their houses with a view to picking up quarrels during the procession on Ashura (pp. 154–5).

At the political level, the support extended by the Shi'a élite to this very expensive festival could have been an effort to counter the Sunni leadership of the *urs* of the saints. After the disintegration of the Mughals, who were Sunnis, the British had reasons to encourage the Shi'as (and Brahmans), who must have, as a ruling class, felt deprived during the previous regime. It may not have been an accident that the Nawabs of Murshidabad, Patna, Lucknow and Rampur were all Shi'as.

However, de Tassy was wrong when he thought that all Syeds were Shi'as or Mo'inuddin Chishti and Wali were Shi'a because they have expressed their veneration for Ali. Islamic mysticism is traced to Hazrat Ali; in the Shahbaz Qalandar song mentioned below (see p. 176). Ali has been given the primary place (*pahla number*) among *sufis*. Even the illiterate porter would cry 'Ya Ali !' when lifting a heavy load. Among other things Ali represents, like Hanuman, devotion and prowess.[14]

De Tassy rounds off his *Memoir* with verses from his favourite poet Mir Taqi and his translator would like to end with a passage from Mrs Hassan Ali, which expresses the poignancy of the human situation symbolized through the suffering woman, a theme completely ignored by the Urdu *ghazal* poet.

I have seen females of rank, with their own hands, place red and green wax lights in front of the Tazia in their halls, on the night of Mayndhie. I was told, in answer to my inquiry, What was meant by the solemn process I had witnessed?—that these ladies had some petition to make, for which they sought the Emaum's intercession at the throne of mercy. The red light was for Hosein, who died in battle; the green for Hasan, who died by poison (vol. I, p. 91).

NOTES

1. The *Memoir* is a reprint in the form of a book of de Tassy's three articles with the same title, which appeared in *Journal asiatique*, 1831. A revised version published in 1869 was later incorporated as a chapter in his *L'Islamisme; d'après le Coran; l'enseignement doctrinal et la pratique* (Paris, 1874).

   At least five editions of his translation of Attar's classic were published during de Tassy's lifetime. Interestingly, the first ever translation into English was not from Persian but from de Tassy's French translation.(*The Conference of the Birds: Mantiq Ut-tair; a philosophical religious poem in prose / Farid Ud-Din Attar; rendered into English from the literal and complete translation of Garcin de Tassy by C. S. Nott; brush drawings by Kate Adamson*, London, Routledge and Kegan Paul, 1961.)

2. 'Contrary to the custom in most other countries,' says Schimmel, ' the 12 Rabi'-ul Auwwal was originally celebrated in India as Bara Wafat, the Prophet's death, and people would spend it in reciting texts concerning Muhammad's excellence ... as time passed, however, the day was interpreted as the Prophet's birthday as was the case in other countries from the Middle Ages' (p. 12).

3. All quotations are from the 'new edition, revised and rearranged, with additions' by William Crooke (Oriental Books Reprint Corporation, New Delhi, 1972).

4. *Doctrine et devoirs de la religion musulmane, tirés textuellement du Coran, suivis de l'eucologe musulman traduit de l'arabe* (Paris, 1826).

5. The *urs* of the *sufi* saints, who de Tassy thinks do not have annual festivals, fall on the dates mentioned against their names: Abdulqadir Gilani (11 Rabi'us sani); Sakhi Sarwar(28 Jamadiul Auwwal); Qutbuddin Bakhtiar Kaki (14 Rabi'ul Auwwal); Baha'uddin Zakaria (7 Rajab); Fariddudin Ganjshakar (5 Moharram); Bu Ali Shah Qalandar (26, Rabi'ul Auwwal); Nizamuddin Awliya (17 Rabi'us sani).As his association with Shaikh Saddo would suggest Shamshuddin Daryai was an apocryphal *pir*.

   For those who would support the hypothesis that the Bahra'ich shrine was originally a sun temple, it would be interesting that in addition to the *mela*, described by de Tassy,

Salar Mas'ud's *urs* is celebrated according to the 'Urdu' month on 12 Rajab. Thus in the oldest Muslim shrine of India, the festival of the death of the hero is separated from the celebration of his birth.

6. A few years later, in his '*Notice sur les fêtes des Hindous, d'après les ouvrages hindoustani (Nouveau journal asiatique*, February 1834, p. 98), de Tassy says that the Hindu months are '*luno-solaire*', without explaining what he means by this word.

7. An example of how deceptive names can be is that of Fa'iz. It is very difficult to accept Sadruddin as his given name (see above, p. 12). In a document we find the names of the three sons of Zabardast Khan alias Ali Mardan Khan (given name, Muhammad Khaleel) who were Hasan Beg, Muhammad Mehdi and Muhammad Taqi. If Zabardast Khan had only three sons, Fa'iz could have been one of them, but which of the three, we may never be able to establish (*Faiz Dehlvi and Divan-i Fa'iz*, by Syed Mas'ud Hasan Rizvi Adeeb, Aligarh, 1965, p. 23 and pp. 25–26).

8. In the Perso-Arabic script vowel signs are not given unless they are necessary. It was therefore natural for de Tassy to confuse between *bira* (folded leaf of pan) with *baira* (ship, fleet.)

   Similarly, de Tassy could not distinguish between the vowel sounds in *bari ye* and *chhoti ye*, since such a distinction was not made in lithographs, so he misqoutes Mir by making *bayan* a feminine; it should have *tere bayan* and not *teri bayan*.

9. Orientalists thought that the *ghazal* was a poem, even when they were at a loss to notice the abrupt transition from one verse to another. William Jones would call the *ghazal* an ode and as his French translation of the *ghazal* of Hafiz would show, the translator has tried to impose some kind of unity on the composition which apparently there was not (see A. J. Arberry, *Classical Persian Literature*, London, p. 337).

10. According to Nicole Balbir de Tugny, the British rulers encouraged, 'for whatever reasons', the growth of Urdu and Hindi along separate lines, till by the 1850s they became two independent languages, and in this connection she refers to the role of W. Price. She also mentions Grierson's repeated assertion that the '*style hindi*' was the gift of the British. She absolves Gilchrist of any responsibility in this regard; the popular language in India for him was Hindustani, whether

it was written in the Devanagri or Persian script, whether the author was a Hindu or a Muslim. According to him,' Hinduwee, Hindooee, or Hindwee 'was the language of India before the Muslims came to the country (see 'De Fort William au Hindi Littéraire: la Transformation de la Khari Boli en Langue Littéraire Moderne au XIXe Siècle,' in Littératures Medievales de l'Inde du Norde, ed. Françoise Mallison, Paris, 1991). This separation could have started when the power of literature as a vehicle for the expression and consolidation of religious identity was recognized. In this connexion, it is interesting to note that de Tassy could in 1834 know an author's religion by the Invocation in a book: if it was *Bismillah ar Rahmanir Rahim*, the author was a Muslim and if it was *Shri Ganesh*, he was a Hindu and if it was *Shri Sarab Dayal*, he was a Sikh (*Les Aventures de Kamrup*, London, 1834, p. 140.). Price was the person who communicated with de Tassy and sent Indian writings to him.

11. It appears that the British rather than the French were the direct beneficiaries of de Tassy's scholarship. As we shall see most of his students were British and he trained some British teachers in Urdu. His *Les Aventures de Kamrup* (*1834*) and *Histoire de la littérature hindoui et hindoustani* (*1839–47*) were published in England under the auspices of the Oriental Translation Committee of Great Britain and Ireland. Interestingly, the manuscript of Abbé J.A Dubois '*Mœurs, Institutions et cérémonies des peuples de l'Inde*', which shows the Brahmans in a very poor light, was bought by the British Governor of Madras and its translation in English appeared in 1816 ; the original was published in France only in 1825.

12. The title of this amateurish, versified account of festivals would appear to be a misnomer, compared to the rich crop of poetry with this name, which flourished in north India during the Middle Ages. Originally, village songs depicting the feelings of a girl waiting for her husband or lover through the twelve months, they were transformed into serious mystical poetry. I would content myself with the following piece from Malik Muhammad Jaisi's *Padmavat:* 'Youth has laden with its oranges the branch of my body, which is defenceless against the parrot of *virah* biting the fruits. Drop suddenly like the spinning pigeon. Your beloved is in the arms of another person, and she cannot escape without your help.

The above is a translation of an extract from the Awadhi classic included by Charlotte Vaudeville in her excellent account, '*Barahmasa: les chansons des douze mois dans les littératures indo-aryennes* (Pondicherry, 1965).

13. 'My Father: his Life and Reminiscences', in S. Khuda Buksh, *Essays, Indian and Islamic* London, 1912, pp. 159–60.

14. In a picture attributed to the Isma'ilis, Hazrat Ali, as the tenth Avatar of Vishnu, is shown riding the white mule, *duldul*, with the monkey god Hanuman serving as his umbrella-bearer (see Schimmel, *Islam in India and Pakistan*, Leiden, 1982, plate XXXIX).

1
Muslim Festivals

PRELIMINARY OBSERVATIONS

THE RELIGION of the Hindus has generally attracted the attention of scholars preoccupied with India, and of travellers who have traversed the beautiful provinces of the country and have brought back the fruits of their observations. But there is not enough material available on the religious practices of Indian Muslims, who, under the Mughals, were for several centuries rulers of the peninsula up to the end of the Ganges, who even now are heads of several states, and constitute a population of 20 millions,[1] a number which is increasing every day. Scholars have talked little about them and have generally ignored the precise nature of the state of their religion and the details of their religious practices. The absence of conclusive information on Indian Muslims has been felt particularly by those who wish to read Hindustani and Persian literature written in India, and who are interested in deciphering the inscriptions on their monuments in that beautiful corner of the earth. Thus, we come across frequent allusions to religious practices, about which no author has given any details and about people whose lives nobody has made available to us. D'Herbelot himself and the writers whose names he has included in his *Oriental Bibliography*, are of no help and therefore one has to rely on other sources. It is in order to fill, in part, this gap that I have undertaken the task which I submit today to the friends of India, in the hope that they will find certain new facts about the esoteric and exoteric doctrines of Indian Muslims,[2] who have been the subject of my interest.

It may be enough for me to indicate below, in a summary form, the principal Hindustani works from which I have drawn the material for the *Memoir*.

1. *Bara Masa*, or the Twelve Months, a didactic poem by Kazim Ali Jawan, the author of the tale of *Shakuntala*. In this poem, which reminds one of Ovid's 'Fasti' and of our own poet, Lemierre, Jawan has given a faithful account of the festivals of Indian Muslims. A very recent work, written a little before its publication at Calcutta in 1812, it presents a picture of the contemporary religion of Indian Muslims. Nothing of this work has yet been translated.

2. *Araish-i Mahfil*[3] or *Statistics and History of Hindustan*, by Mir Shair Ali Afsos, who has made an elegant translation of Sa'di's *Gulistan*, *Pindnama* etc., and has published his valuable *Diwan*, a copy of which is to be found in the East India Company Library, London. Only the first part of the *Araish-i Mahfil* was printed in Calcutta in 1808, the entire manuscript lying at present in the Old Fort William College.[4] In the first part of the book Afsos describes places, not leaving out of the account descriptions of eminent people who lived there or are buried there. This is how we get brief accounts of important saints who are venerated by Indian Muslims. His description

is faithful and reliable because he is clear-headed and free from the prejudices which often blind his co-religionists. He begins with a warning in his Preface that he has spoken of most of the saints only because he had to follow the book he had used as the basis of his own work.[5] 'The two worlds (the present and the future) will be so full of saints that I would recognize as my patron only Ali, God's elect',[6] he says.

What makes his account reliable is the fact that several persons[7] he describes were his contemporaries and some of them were known to him.

With the exception of a few extracts, which have no relevance to the theme of this *Memoir* and which have already been published by me in *Journal asiatique*, nothing from this important work has been translated.

3. *Diwan-i Wali* [8] or collected poetry of Shah Waliullah, the Father of Hindustani poetry, *Babae Raikhta*,[9] as he is affectionately called by his people. Wali was from Gujarat and he lived in the second half of the seventeenth century. His *Diwan* is the equivalent of that of Motanabbi in Arabic, of Hafiz in Persian and of Baqi in Turkish. His work has not yet been translated and what is surprising is that it has not yet been published, even as much less important works in Hindustani have come out from Calcutta and elsewhere.[10] It was this *Diwan* which inflamed the poetic ardour in Afsos and which inspired him to write in his mother tongue,[11] to which many of his compatriots preferred a language dead to them; much in the same way as in Europe, Latin once usurped the rights of the national languages.

4. *Diwan-i Fa'iz*, or the collected poetic works of Muhammad Sadruddin, whose sobriquet, *takhallus*,

was *Fa'iz*. This work also remains unpublished and nothing of it has yet been translated.

5. *Hidayat-ul Islam*,[12] or a Muslim eclogue in Arabic, Persian and Hindustani, printed in Calcutta in 1804. It is the same work, a translation of which I included in my *Doctrine et devoirs de la religion musulmane*, in 1826. I had then omitted the *fatihas*[13] for the Muslim saints of India; but they are now useful for this *Memoir*.

6. *Gul-i maghfirat*[14] by Mir Haidar Bakhsh Haidari, which is the history of the Muslim martyrs up to the death of Husain in Karbala. It was printed in Calcutta in 1812, and has not been translated.

7. A collection of Hindustani proverbs, comprising the second part of the excellent work, *A Collection of Proverbs and Proverbial Phrases in the Persian and Hindustani Languages*, published in Calcutta by the famous Indologist, H. H. Wilson. These proverbs have been collected by the Oriental scholar, Thomas Roebuck, who was a friend and collaborator of Dr Gilchrist. Each proverb is translated faithfully and has an interesting note on it.

I do not list here works which I have cited only in passing, particularly Mir Hasan's poem, *Sihrul Bayan*,[15] which is a masterpiece of imagination and taste, and which is a significant work of Hindustani literature. The other poems not listed here are by Mir Taqi, whom I introduced a few years ago, by publishing some of his pieces.

I am thus going to describe with reference to the books mentioned above, the festivals of Indian Muslims, which could also be observed in Persia or even in the whole of the Muslim world, but which have in India acquired special rituals. I shall speak of some superstitious practices which have been introduced on account of the contact with Hindus. In the end I will give biographical notes on some Muslim saints who are popular in certain parts of the country but unknown outside India, some of these saints being venerated equally by Hindus and Muslims.

What strikes me most about the religious ceremonies of Indian Muslims is the innovations which make them appear as local phenomena. Established as they are on account of unconscious Hindu influences, there are wholly new ceremonies, which conform little to the spirit of the Quran and are sometimes even contrary to its spirit. Further, in many cases, Muslims make pilgrimages to the tombs of saints, some of whom are actually non-Muslim, and perform there semi-pagan ceremonies.

As a matter of fact, the ceremonies prescribed by Muhammad appeared too simple in a country which was dominated by an allegorical and idolatrous religion, one which appealed to the senses and the imagination rather than to the spirit and the heart. Muslim festivals in India are charged with pagan ceremonies and have taken on a lavish appearance. Muslim pilgrims in India are not affected by the austerity which characterizes the pilgrimage to Mecca and Medina. Indeed, one would imagine that they are completely Hindu festivals. [16]

Among the Sunni Muslims there are only two important festivals, one at the end of the month of fasting, Ramazan, that is, Id fitr, and the other, that of the victims, Id qurban, also called in India, festival of the Ox, Baqr Id or simply, Id, which is celebrated in memory of the sacrifice of Ishmael.[17] The Shi'as have a few more, yet not enough in a country which is used to a surfeit of Hindu festivals. Indian Muslims have therefore created new ones, which both Shi'as and Sunnis are only too eager to celebrate. Some of them are consecrated to the memory of *pirs*, who are to Muslims what *deotas* are to the Hindus. They visit the tombs of these *pirs* on Thursdays and sometimes on Fridays.[18]

The Muslims festivals, which I am going to describe soon, appear to read like those of the Hindus. To give an example, the solemnity of *ta'zia*, mourning, observed to commemorate the martyrdom of Husain, is comparable, on a number of points, to that of the Durga Puja, which the Hindus celebrate in the month of Katik (October–November) in honour of Durga, goddess of Death and the wife of Shiva or Mahadeo. Like the Durga Puja, the *ta'zia* is observed for ten days. On the final day the Hindus immerse the image of the goddess in a river amidst huge crowds and great pomp, while a thousand musical instruments are played.[19] The same thing happens with the Muslim festival. Mourning is observed for ten days and the *ta'zia*, a replica of the tomb of Husain, is generally immersed in a river [20] with the same pomp. One will notice in the following account how much this and other Muslim festivals have borrowed from Indian practices. The noisy processions remind one of those of Jaggannath[21] and of other pagodas, where, hardly edifying but indispensable troops of bayadères[22] and prostitutes join the religious procession. The offerings made by Muslims to these saints are the same as those of the Hindus: rice, ghee and flowers.

Among the festivals I am going to talk about, there are some that have never been described before. Some of them have been taken note of mainly by Chardin, but I need not take them up,[23] either because these are not known in India or because I have not found them mentioned in the Hindustani books which I have been able to consult. However, the festivals described by Chardin and other writers are distinguished, as I have said, by specific practices and ceremonies, and so I realized it is necessary to mention them.

Indian tolerance has diminished the fanaticism of the Muslims: Sunnis and Shi'as do not have in India the animosity that divides the Turks from the Persians. Ordinarily, they live in amity and, except in certain cases, they participate equally in festivals.

It is not necessary to dwell here at length upon the two principal sects into which Muslims are divided. One may compare the former with Catholics and the latter with Protestants, and not the other way, as suggested by the late Mr Langlès in the note that he adds to his edition of Chardin.[24] Indian Muslims are divided into two sects but as I have just said there is no animosity among them.[25] Some Muslims are at the same time both Shi'a and Sunni. Thus the famous poet, Wali, praises the first four Caliphs, Abu Bakr, Umar, Usman and Ali in a few words, and then at length and emphatically, Ali and his sons, Hasan and Husain, whom he calls the Imams of this world, *imamé jahan*.

Each of the two sects has its own honoured saints who are respected equally by the other.

The Muslims of Ali's sect are generally called Imami, that is, supporters of the Imams, rather than Shi'as or dissidents. They are also called, *Ali mardan*,

Ali's men[26] and *haidari* from the Arabic, *haidar*, lion of God, *asadullah*.

In connection with practices borrowed in minute detail from Indian sources, mention may be made of the ridiculous devotion they sometimes have for apocryphal monuments and fantastic relics. For example, near the city of Oudh, there are two large tombs, each 7–8 *gaz*[27] long, believed to be the tombs of Seth and Job,[28] where people throng on Thursdays to recite the *fatiha*.[29] Such also is the tomb of Lamech or Lamag, Noah's father, which is said to be at Ali Chang, a village in Kabul, from which is derived Lamagan,[30] the district in which this village is situated. Such also is the supposed impression of the Prophet's foot, *qadam sharif*, which is to be found near Banaras, not far from Aurangzeb's palace and the pond called Batsha Mochan: people of all classes assemble here on Thursdays.[31] Such also is the beautiful but ridiculous monument of Cuddapah[32] built in 1135 H., 1723 AD to preserve the hair of the Prophet's beard, which is kept in a casket of gold.[33]

One of the most remarkable practices of the Indian Muslims on which it is necessary to dwell a little, is the external signs of veneration which the people show to the saints, who are called *pir*, or *wali*, in Hindustani. As we have already said, they are for Muslims what the numerous gods are for the Hindus. In every town, in every village, even in the capital of pagan India, Banaras,[34] we find tombs of these saints who are, however, unknown elsewhere. Some of these saints gave their names to the cities which, in course of time, developed round their tombs. Qutubuddin has given his name to the town Qutub or Qutb in the province of Delhi[35] and Hasan Abdal, a famous, pious

Muslim, to the beautiful valley in the province of Lahore and to a town where he is laid to rest.[36] Similarly, Rauzah, or tomb, is a town in Aurangabad, famous for the tombs of celebrated saints.[37]

Some of these *pirs* are very well-known and festivals are held for some of them throughout the country. However, the saints about whom I shall talk in detail later are only six in number: Khwaja Khizr, considered usually to be Prophet Elijah and the five saints who, I believe, are some *pirs*, whose devotees call themselves *panch piriya*,[38] or the devotees of five *pirs*. These *pirs* are so popular that their names are given to the lunar months in which their feasts fall. No particular feasts are associated with *pirs* other than these six though some of them could be equally venerated. It is necessary, however, to mention them in order to make our account more useful and complete. However, it is not possible to deal with all those *pirs* whose popularity is only local. To prepare an account of these local *pirs* will require a lot of work and the result will not be commensurate with the toil involved. It will be of little use enumerating a host of saints, more or less obscure, with the legends associated with them being hardly believable. Colonel Brigg came to the same conclusion in his new translation of Farishta. He deleted in his translation chapters concerning the Muslim saints of India as he believed they were of little interest to the European reader. For a work of this kind, it would be necessary to have recourse to sources other than the Hindustani works which I have consulted for this *Memoir* and also Farishta's account which ends at the year 1611. Moreover, since Farishta, who wrote more than two hundred years ago, new names have been entered in the diptyque of the Muslim community. Even after Englishmen took over the sceptre of power, many have distinguished themselves by their piety and the English, just appraisers of merit, have not always succeeded in engaging them into their service. Such a person was Maulana Abul Khair of Jaunpur who came of the Faruqi family and was a Hanafi. This saintly person declined Governor W. Hastings' offer of a place in the court of Banaras. Afsos says,[39] 'Having resolved to turn his face from this world, he knew how to be content with his lot and he left the corner of his retreat and set out to enjoy the plenitude of immortal joy in 1198 [1783-4].'

I might, at a later stage give a complete biography of these saints, taking into account the new researches done on them, but for today I have imposed on myself, strict limits.

I have already said that among the saints venerated by Muslims there were a few who professed the cults of the Vedas. In the same way many Muslim saints are venerated by the Hindus.[40] One such saint is Shah Lohani at whose tomb in Monghyr both Hindus and Muslims come to offer oblations, particularly during the marriage season and other solemn occasions.[41] Equally venerated by the two communities is the tomb of Shah Arzan (d. 1032, 1623),[42] situated in the western suburbs of the European station of Patna.

This reciprocal tolerance has its source in a broad view of life, which one would not have expected from Muslims but which, nonetheless, is in complete conformity with the spirit of the Quran. According to Muhammad, there is only one true religion and God has revealed this fact through prophets and saints. Thus Moses, Jesus Christ, Zoroaster[43] and Brahma have, according to His system, preached the same doctrines. But men could not comprehend this and they altered the divine cult and it was in order to restore the original purity of religion that Muhammad was sent. One can see, therefore, that it is not un-natural that the Muslims venerate saints who are out-side their religion.

Among the Hindus revered by the believers in the Quran mention may be made of Baba Lal and Kabir, about whom we shall talk in the second part of this *Memoir*.

The low-born Muslims, not content with honour-ing some Hindu saints, often join the pagan fêtes of the Brahamanic religion and go as far as to offer obla-tions to idols.[44]

Among the Muslim saints many were outwardly licentious, in the same way as was the famous Persian poet Hafiz, who is known to every Orientalist for his erotic mysticism, but is at the same time a respected Sufi, whose tomb near Shiraz draws a large number of devotees.[45]

The titles we have given to these saints lead us to another observation. In India there are four classes of Muslims: the Syed, or descendants of Muhammad through Husain; the Shaikh, or the Arabs vulgarly called Moors;[46] the Pathan or the Afghan and lastly the Mughal. All these four classes have produced saints, who can be identified with their class denominations and others especially conferred on them. Thus, Syeds are called Mir; Pathans, Khan; the Mughals, Mirza, Baig, Agha and Khwaja. The word Shah or Sultan, literally, a man endowed with supreme power, is often employed as an honorific before the given names of these *pirs*, maybe because they are considered kings of their souls and masters of their passions.[47] Independently of these titles, their names are divided into three parts: the given name or *alam*, such as Muhammad, Ali, Husain etc.; the honorific *laqab*, such as Saifuddaulah, (the Sword of the Empire), Asafjah, (the person who has the dignity of Asaf, Solomon's minister), etc.; lastly, the nickname, which one chooses for oneself, *takhallus*, appropriation.[48] This appropriation is ordinarily an abstract noun such as *tapish* (affliction), *qudrat* (power) etc. In place of the poetic appropriation, which a poet would never forget to adopt, many a saint is distinguished by his patronymic *padbi* [sic] which is shared by all those who belong to his spiritual order. Such a patrynomic is Chishti, about which we will talk later. Each *pir* belongs to a well-known spiritual lineage, and while initiating him to contemplation, he gives to each of his disciples, a genealogical tree *shajar nama*[49] of the persons who comprise that spiritual lineage. And each spiritual family is founded as a monastic order with a superior or head, *mansnad nashin* or *sajjada nashin*. The succession to this headship takes place with the handing over of the stick and the robe of the deceased chief.[50]

Pir means literally an old man, but in this context, a person who has spiritual dignity, much like the guru of the Hindus. Muslims who dedicate themselves to the study of religion and practise piety, are required to choose a spiritual guide. Wali has said, 'Follow the *pir* as his shadow does the man.'[51] Many of the *pirs* are venerated as saints after their death. That is why the word *wali* is synonymous with *pir*, and signifies a saint as much as the word *wali* does.

In times of distress people turn to these *pirs*, who during their life, would help them by praying to God on their behalf. Sometimes people go to them for amulets, *ta'wīz*.[52] Hindus as well as Muslims believe that tigers and leopards are the property of these *pirs*, that is why the natives resent the Europeans hunting tigers.[53] In the delta of the Ganges, called Sunderbans, one may come across Muslim holy men who claim to possess charms against the cruelty of tigers. These people live in miserable huts on the rivers and are greatly respected[54] by passers-by, Hindu, as well as Muslim, who give these holy men food and cowries to make them propitious.[55]

Tombs have different shapes, which it is not useful to describe here. However, most of them consist of a small chapel in the middle of which is the saint's shrine. Sometimes it is built on a raised platform without steps, so that people are obliged to recite the *fatiha* from a distance,[56] and they may not come too close to it. The tombs of saints are called variously, *dargāh*, shrine, *mazār*, place of pilgrimage, *rauzah*, garden. These three words always refer to places where saints rest, while *turbat, maqbara*, etc., are the tombs of people who are not objects of religious reverence. From *rauzah*, garden, in the sense of tomb, is coined the composite word *rauzah khwan*, which means a person or persons whose profession is to recite the Quran and offer prayer at the saint's tomb; this word refers especially to those who recite praises of Husain during Moharram.

Homage is paid to the saints by devotees who go in a procession to their tombs on certain solemn days, generally on Thursdays and sometimes on Fridays too[57], and offer prayers and oblations at the tomb. In these processions they generally carry pikes called variously, *chhari*, stick, *naiza*, lance and *jhanda*, banner, since they attach to them pieces of cloth which make them look like flags.[58] On arriving at the tomb, they hoist into the ground these pikes which they take out when they return, These processions are called *maidni*, and in certain cases *chhari*,[59] and are led by *faqirs*.

The offerings they place at the graves of saints consist mostly of flowers, sweets, pastry, and sometimes *mash dal* (vesces),[60] mustard oil and molasses.[61]

They make these offerings in mosques also; Hasan has said, *He offered oblations in the mosque.*[62]

This offering is called *fatiha*, an Arabic word, which means literally overture and refers to the first chapter of the Quran. From that the word has come to mean the prayer offered to the saint, the recitation of the first chapter, which follows it and finally the words of prayer which accompany the offering.[63] However, these prayers are not exactly made to the saints, and they can better be compared to the collects of the mass of Catholic fêtes celebrated in honour of saints to whom they do not pray directly. In other words, although they have great veneration for the saints, it cannot be said that the Indian Muslims pray to them.

In case the *mulla*, the priest attached to the tomb, is asked to make the offering himself, the payment made to him is called, *chiraghi*, the gift for the expenses for the light, *chiragh*.[64]

Endowments made to enrich the shrines are called, *nazre aiyima*,[65] presents in the memory of the Imams. Rich landlords consider it their duty not only to make endowments of necessary land for building the tombs of saints and for the accommodation of crowds of pilgrims but also to permit *melas* or fairs being held on

their land. Moreover, as an act of piety, they endow
land, the revenue of which is spent on the raising and
maintenance of the building and the employees' salaries;
and on making provisions for lighting the place. These
pious gifts are called *piran* and *nazre dargah*.[66]

A *mela* is not exactly our fair. This is the name
given to the gathering of pilgrims and merchants
drawn by devotion or monetary interest and often for
both, to the place considered sacred, on the occasion
of the fêtes of some Indian gods in the case of Hindus,
and of certain Muslims who are considered saints. In
order to meet the demands of the multitude gathered
there, merchants set up shops.[67] Thus the word *mela*,
fair, is often mixed up with pilgrimage, *ziyarat*, among
the Muslims and *tirath* among the Hindus.[68] In addition to those attracted by religious devotion or business interests, many go there merely for the sake of
curiosity or for pleasure and finally thieves and
swindlers sneak there in the hope of exercising their
special art. Thus, these gatherings comprise the *faqirs*,
devotees of all classes, musicians and jugglers, prostitutes and dancing girls, magicians and libertines,
rascals and thieves.[69] The following account of a festival,[70] half-mundane, half-religious, can give us an
exact idea of these *melas*. This is about the fair which
is held annually on the first Sunday of *Jaith* (May–
June) at Bahra'ich, in the kingdom of Oudh near the
tomb of the famous Muslim martyr, Salar Mas'ud
Ghazi, about whom we shall talk in greater detail, in
Part I of the *Memoir*.

This annual fair is held in the middle of a forest,
which is now free from wild beasts. A thousand
scenes offer themselves to the sight. One sees

swings everywhere; with each tree a balancer is suspended. Tents and merchants' stalls are erected all around; sweets of all kinds and colours are artistically spread out; bread of different varieties, made with milk or only with water, cover the bakers' table; while at other places meat roasted in several ways is placed on plates. Rice cooked in different styles and heaps of fresh and dried fruit are offered to the buyer. There is great demand for *paan*, which is sold in packs of a hundred leaves and small boats called *bera*,[71] of flowers which the devotee buys in order to offer them to the saint on his wishes being granted.

There are musicians who play on different instruments, jugglers who play skilful tricks, dancers from the Deccan with astonishing suppleness. Graceful bayadères and bold rope dancers are the most remarkable among them. In the middle of the ravishing spectacle, intoxicating liquor made of the exudation of hemp flowers[72] is passed on everywhere. Beyond themselves, the boozers start shouting *haai* (alas!) and *hu* (God!) However, everyone goes to the tomb and pays homage by offering flowers and sweets. Singers and musicians express their devotion in their own way. Thousands of candles and lanterns shed their brightness on the flowers of lotus and cypress. This continues from morning till sunset. Then the pilgrims return home. Awaiting their return impatiently, friends and relations surround them. Coins and flowers are showered on them and everybody wants to kiss their feet. And they can get away from the crowd only after distributing among them things which had touched the saint's tomb.

NOTES

1. Hamilton, *East-India Gazetteer*, vol. I, p. 648.

2. The exoteric, from two works, first, *Exposition de la foi musulmane* and, the second, *Doctrine et devoirs de la religion musulmane* and *Eucologe musulman*. The esoteric, with the publication of the moral, or rather, the mystical allegories of Azizuddin el Moqaddasi's *The Birds and the Flowers*.

3. *Araish-i Mahfil*, or The *Ornament of the Assembly*.

4. Abolished sometime ago, as a measure of economy.

5. Afsos alludes to *Khulasatut twarikh* or *al hind*, of which his work is far from being a servile translation. See my comments on this book in my *Rudimens*, p. 16, and in *Journal asiatique*, vol. viii, p. 239 sqq. At the time when I was drafting the latter article, I did not have with me this book, a manuscript of which I have since purchased. That is why I based my comments on the notes I had made; otherwise I would not have said that it extends up to the death of Aurangzeb: as a matter of fact, it ends at 1069 (1659), the year when his brother, Dara Shikoh was taken prisoner.

6. *gar do ālam pur az wali bāshad pire mā murtazā ali bāshād. Araish-i Mahfil*, p. 5. By this verse alone, which is in Persian, and apparently a citation, one can conclude that Afsos was a Shi'a.

7. Some of them, for example, are Shah Qutubuddin of Allahabad, as famous for his poetry as for his piety (*Araish-i Mahfil*. p. 82): Kamal Shah Muhammad Afzal of the same city, as famous as meditator as the author of two *diwans*, or collection of poetry, one in Persian and the other in Hindustani (ibid., p. 83): Maulwi Roshan Ali, a person as religious as scholarly, then head professor of Arabic at Fort William College (ibid., p. 93).

8. A *diwan* is, as a matter of fact, a collection of *ghazals*, whose rhymes cover, gradually, all the letters of the alphabet. The *ghazal* is a short poem, which can better be compared to the Italian sonnet. It is composed of six to twelve verses, which have the same rhyme. Its theme is ordinarily erotic, but most often physical love is only a veil to conceal the spiritual love, which is the well-known topic of all Muslim poets. In the final verse of the *ghazal*, the poet introduces his name very skilfully

and this is where lies the difficulty and the charm of this form of composition.

9. Gilchrist, *Hindoostanee Philology*, p. 484.

10. I had the opportunity of buying two good copies of this *Diwan*, and this might lead to my determination to get it published.

11. Preface to *Bagh-i Urdu*, translation of the *Gulistan*, p. 14.

12. *Hidayatul Islam*, literally, *Guide to Islamism*.

13. We shall find below an explanation of this word.

14. *Gul-i maghfirat*, literally, *The Rose of Forgiveness*.

15. *Sahrul beyan*, that is, *The Magic of Eloquence*.

16. *Araish-i Mahfil*, pp. 179, 180.

17. According to the Muslims, as we know, it was Ishmael and not Isaac, whom Abraham wanted to sacrifice.

18. Every Friday, many elegant young men go to the tomb of Pir Jalil, near Lucknow, to stroll and to enjoy themselves, while devotion attracts a large number of low-born people, who offer *khichri* (peas and rice boiled together) vesces and bitter oil. *Araish-i Mahfil*, p. 100.

19. *Araish-i Mahfil*, p. 133.

20. Shakespear, *Dictionary*, p. 251.

21. A temple built 4000 years ago by Raja Indra Sain, in the city of Parsotam, province of Orissa. *Araish-i Mahfil*, p. 133.

22. The word we have adopted in our language is Portugese *bailadeira*. These female dancers have many names in Hindustani; the most common being, *Ramjani*, that is, gentle; *nautchi*, woman dancer, and *kanchani*, which is most common and which, according to Bernier (*Voyages*, vol. XI, p. 59) means golden from *kanchan*, gold.

23. Such is the ridiculous ceremony of *sar wa tan*, the head and the body, performed on Safar 20, in memory of the so-called miracle which happened to Ali, when, according to the Shi'as, his head joined his body forty days after it was severed. See *Les voyages de Chardin*, ed. by Langlès, vol. IX, p. 67.

24. Vol. vi, p. 173.

25. However, as a measure of precaution, the police requires the Sunnis not to stir out of their houses during Moharram, fearing that some Shi'as, either on account of fanaticism or under the influence of liquor may become violent against them. *Asiatic Journal*, xxvii, p. 255.

26. *Voyages de Bernier*, vol. I, p. 14.

27. Measure of length equal to three feet.

28. There is another tomb of Job near Huleh, a town on the bank of the Euphrates. See Langlès, *Voyages de l'Inde à la Mecque* by Abdul Karim, p. 126.

29. *Araish-i Mahfil*, p. 95.

30. Ibid., p. 205. The author of *Ain-i Akbari* says the same thing, but his name is also pronounced as Lagman, which demolishes the supposed etymology. From this is derived Lagaman, which refers to the local language of the district. See Hamilton, *East-India Gazetteer*, vol. II, p. 133.

31. *Araish-i Mahfil*, p. 88. Another impression of Muhammad's foot is found in the town of Cuttack. Engraved on a piece of stone, it was brought from Mecca, and is kept in an octogonal shrine. Near Naraingunj in Bengal, there is a third impression of the Arab prophet's foot, very much venerated by Muslim devotees, who come in great numbers from Dacca and other adjoining towns. A fifth impression has made a mosque in Gaur, famous; lastly, such vestiges as fabulous as others, are not rare in other places in India. Hamilton, *East-India Gazetteer* vol. I, p. 472; vol. II, p. 292.

32. In the province of Balaghat.

33. The casket has a crystal cover, with small holes through which water is passed once a year on a specially solemn occasion, when pilgrims from all places come to see the relic.

    Muhammad had the habit of passing his hand through the beard, when in informal company. When one of the hairs broke, the disciples grabbed it and kept it carefully. Such is the origin of the relic. When the famous Haidar conquered Kudappah, he took hold of the hair and took it to Seringapatnam, where it was kept till the English captured the city. It is not known what happened to the relic after this event. Skinner, Note, *Asiatic Journal*, N. S., vol. II, p. 328.

34. We shall find further down the life of a saint buried in this town.

35. Hamilton, *East-India Gazetteer*, vol. I, p. 473. See below the article on this saint.

36. Ibid., p. 672.

37. Ibid., vol. II, p. 471.

38. Shakespear, *Dictionary*, p. 205. According to Colonel Harriot (*On the Oriental Origin of the Gypsey, Transaction of the Royal Bengal Asiatic Soc.*, vol. II, p. 530), Muslims in India call *panchpiri*, a class of wandering tribes who have some resemblance with our Bohemians.

39. *Araish-i Mahfil*, p. 93.

40. Hamilton, *East-India Gazetteer*, vol. I, p. 648; *Asiatic Researches*, vol. XVI, p. 135.

41. Hamilton, *East-India Gazetteer*, vol. II, p. 237.

42. Ibid., vol. II, p. 382.

43. It is said that one of the twelve Imams made this remark in connection with Zoroaster: *kana nabian o hakiman*, he was a prophet or at any rate a wise man.

44. Hamilton, *East-India Gazetteer*, vol. I, p. 648.

45. One may mention in this connection Maulwi Mir Askari, who was a descendant of Husain and was an Imami. It is said that externally, this esteemed person was without any self-restraint, but internally given to religious contemplation. He had many disciples whom he taught spiritual science and who acquired perfection through his company. He died in 1190 (1776–7) in Jaunpur, where one can see his tomb, which is a place of pilgrimage (*Araish-i Mahfil*, p. 93).

   The date of his death is calculated, says Afsos, from the words *bardullah mazjam'a* [sic], may God freshen the place, where he rests. By adding up the numerical value of each letter, which makes up this chronogram, one gets the date mentioned above.

46. The Muslim Arabs, who settled, during the times of Caliph Valid on the Malabar coast and in north India, are even today called Maures (Moors): Pathans, or as they are also called Afghans, have nothing in common with these Arabs, except their

religion. J. R. Forster, *Notes sur le voyage aux Indes Orientales du P. Paulin de St.-Barthelemy*, vol. III p. 133.

47. Hamilton, *East-India Gazetteer*, vol. III, p. 271.

48. It could also be possible that this denomination is derived as M.Belfour says, from the convention, according to which, the poets place the sobriquet at the end, *khala*, of the poetic composition. See *The Life of Ali Hazin*, p. 21.

49. Shakespear, *Dictionary*, p. 544.

50. One finds in IV [*sic*], Kings ....II, 13, that Elijah was entrusted with Elisha's mantle, when he left.

51. *sāyā naman tu pir ki dayam dumbāl chal.*

52. M. Reinaud, who has recently published a very useful book on Muslim monuments has sent me the sketch of one such amulet, which in India is given to the mother of an infant, and which she should wear on the right arm. One can read there, along with certain versets of the Quran, names of certain Indian Muslim saints, who have acquired various degrees of fame: Mo'inuddin, Kabir, Qutubuddin, Fariduddin, and Nizamuddin, saints on whom we have notes in this *Memoir*.

53. The reason is perhaps that tigers are useful where there are *jungles*, that is, trees and tall grass. They destroy wild dogs and deer, so feared by farmers, and these tigers return when they have purged the country of these dogs and deer. Hamilton, *East-India Gazetteer*, vol. II, p. 605.

54. *kauri*, shells used as money.

55. Hamilton, *East-India Gazetteer*, vol. II, p. 605.

56. *Araish-i Mahfil*, p. 100.

57. Ibid., p. 110.

58. In Egypt they use in similar ceremonies, leafless branches of palm trees, called *ma'qrea* or *tartaqa*.

59. See the article on Madar.

60. *māsh, phaseolus max or radiatus.*

61. In this connexion, Afsos seeks permission to make the observation that while admitting the fact that the saints to whom the offerings are made had revelations and the power of miracles, yet one cannot but acknowledge the fact that they

had very bad taste, if one believes not only that they accept such offerings after death, but also that they desire these offerings. *Araish-i Mahfil*, p. 100.

62. *lagā āp masjid men rakhne diyā, Sihrul biyān*, p. 27.

63. In my *Eucologe musulman*, p. 215 et sqq, I have included translations of a number of prayers called *fatiha*, and there are a few more in this *Memoir*.

64. Shakespear, *Dictionary*, p. 330.

65. Ibid., p. 93—Rousseau, *Dictionary of Mohammedan Law*, p. 181.

66. Rousseau, *Dictionary of Mohammedan Law*, pp. 180, 184; Shakespear, *Dictionary*, p. 224.

67. Hamilton, *East-India Gazetteer*, vol. I, p. 187.

68. As we have seen above, Muslim pilgrimage has many similarities with that of the Hindus and is often even identical.

69. *Araish-i Mahfil*, pp. 100, 111, etc. Hamilton, *East-India Gazetteer*, vol. I, p. 231.

70. This is an extract from *Bara Masa*, p. 50 et sqq.

71. *bira*. These small boats are floated in rivers by Muslims in honour of Khwaja Khizr. See the article on the fête of this prophet.

72. More generally called *bang* (F. Gladwin, *Materia Medica*, n.74) For the use of hemp as a drink see, *Arab Chrestomathie* by M. le Baron S. de Sacy, vol. I.—It appears that the devotees of Madar and probably of Salar Mas'ud also, make extensive use of this liquor. See *Asiatic Journal*, N. S., vol. IV, 75.

PART ONE

Lunar Fêtes

The Month of Moharram

Ceremonies on Husain's martyrdom Of all the grand solemn occasions of Indian Muslims, it is this festival which is observed with the greatest pomp and show. It is the most important festival of the Shi'as or Imamis, although it is celebrated by others too. It is generally called Moharam, because of the lunar month in which it falls, or more often *dahā* or *dāhā*, a Persian word derived from *dah* or ten, from which also the moon of *daha, dahai ka chand*.[1] This festival is also called *ashura* or *ushra* derived from *ashr*, which in Arabic, like the Persian *dah*, means ten. These terms owe their origin to the fact that Husain's fête lasts for ten days, that is to say, the first ten days of the month. The tenth and final day is the anniversary of the death of the blessed Imam, which took place on the same date in 61 H. (10th October 680). Jawan cries out in anguish,[2] 'To the Last Day the sorrow of this cruel event will continue to penetrate the hearts of all Muslims.'

Husain and his elder brother Hasan were children of Ali and Muhammad's daughter, Fatima. The events leading to Husain's death are narrated in detail in *Gul-i Maghfirat* (pp. 201–47), and in all books dealing with the advent of Islam. It is not necessary to go into these details which are outside the scope of this *Memoir*. I will confine myself to the main outlines given by Jawan.[3]

Reposing rather too much confidence in the inhabitants of Kufa, who had invited him to go there and become their leader, Husain left Medina. Soon, without any protection, he found himself exposed on all sides to traitors and assassins.[4] Only 72 persons, most of his family and all his cherished friends, remained loyal to him. Surrounded on all sides in the plain of Karbala,[5] they remained there for three days without food; they became very weak and were forced finally to give up their lives. Horrible plight! How can one state the terrible state in which the women in Husain's harem found themselves? They had no way out but to cry and along with tears inundating them, their souls came out of their bodies. At last the sword of violence killed Husain and all his unfortunate companions, except his son Ali, who was sick.[6] The tent of Husain, this angel from the heaven, was pillaged and set on fire; his women were subjected to all kinds of insult and ill-treatment. Full of shame, the sun and the moon turned their glance away from this shattering scene.

To commemorate this tragic event the mournful festival of Moharram is established. Whoever participates in this festival, will find recompense in Heaven. One should express in tears and lamentation what one feels about the cowardly attack which took the life of the Prophet's grandson, and if a person cannot do it himself, somebody should be appointed to do it on his behalf. From the moment that the New Moon appears on the horizon, the devout Muslim begins sighing and moaning and preparing what is called the feast of mourning, *bazm matam ka*; that is to say, providing water to the thirsty and sherbet to mourners. These offerings are prepared daily from the first to the tenth of the month. Everybody, dressed in black, cries, beats the head, plants banners and carries replicas of Husain's tomb.[7] A room with black hangings is decorated, and a chair is placed on a raised platform. It is from there that one reads out for ten days the sad story which is the object of this festival. The person who is assigned this task[8] accompanies the reading with groaning that passes all limits. On their part, the listeners express their sorrow by lamentation and cries of *Salām*.[9] Then an elegiac poem is recited in honour of the saint. This poem is full of pathetic details of his martyrdom, and this excites fresh tears in the assembly.

There is another ceremony which I should mention and indicate its purpose. It is said that in order to fulfil the last wish of his brother, Husain wanted to install the former's son, Qasim, as the successor to the Imamat. He put on him the robe of Muhammad's son-in-law (Ali) and repeated the oath of loyalty. In order to commemorate this, the mourners repeat this ceremony during the anniversary of Husain's death.[10]

One can see from the above that travellers who have said that this festival commemorates the death both of Hasan, Ali's elder son, and Husain, the younger one, are mistaken.[11] This solemn ceremony is founded to mark the death or as Muslims would put it, the martyrdom of Husain, and it is only incidental, if in their exclamations the devotees add Hasan's to Husain's name; it is natural that the tragic end of Husain would remind one of almost similar end of Hasan. Although much less solemn, the mourning for

Hasan is observed on his death anniversary on 28th Safar.

Moharram continues for ten days, because Husain was harried for ten days.[12]

We have seen already that pikes and banners characterize Muslim processions in India. The pikes and banners, used at Moharram have a special name, *shadā*.[13] They are hoisted into the ground, around the assembly of mourners, in the same way as at a place of pilgrimage.

The representations of Husain's tomb, or more exactly the shrine in which his tomb is situated, are often decorated richly. These structures are given the metaphoric name *ta'zia*, mourning,[14] or simply *tabut*,[15] coffin. On the tenth day they are carried through the streets and finally buried or immersed in a river or a pond.[16] If these cenotaphs are costly, only the image of the tomb is dispensed with, and the structure is brought back and put in the Imambara or sometimes in the monument of Karbala.[17] In order to re-enact the burial of Husain, they sometimes bury only the flowers from the cenotaph and with this ceremony, end the ten days of mourning.[18]

The room with black hangings, which we have mentioned earlier, is, undoubtedly, the edifice called appropriately, Imambara, a compound word made of the Arabic, *imam*[19] and the Hindustani, *bara*, which, in this context, means house.[20] This edifice is called, *ta'zia khana*, house of mourning.[21] It is found only in India and is made especially to observe the funerary fête instituted in memory of Husain's martyrdom. Afsos tells us that there are a very large number of Imambaras in Calcutta. 'Muslims, both men and women, even of modest means [says he], have in their houses an adjoining room with a small cenotaph,[22] two or three cubit *hath* high, on something like a terrace, *chabutra,* of the same length as its breadth. It is often enclosed and has three other smaller cenotaphs. The enormous cost involved in its construction worries the owner little.'[23]

Moreover, sometimes people are also buried in these Imambaras. In Lucknow, Asafuddaulah is buried in his Imambara,[24] and the Mughal, Baqir Khan, in one he built for himself in the jewellers' quarters.[25] Sometimes, Imambaras are built primarily as tombs for a family.[26] As we have seen above, it is in the Imambaras that the faithful, mostly clad in green or black[27] come to listen to the tragic history of Husain's martyrdom, on the first ten days of the month. To this narrative, delivered from a chair placed on a raised platform, is often added the history of the death of Hasan and other saints. This narrative is, as we have already said, read out in a special tone and with appropriate gestures so as to arouse the listeners' emotions. At every pause the listeners beat their breasts and pronounce alternately the names of Hasan and Husain.[28] Excited by these lectures, bands of devotees rush into the streets making demonstrations of their sorrow and, as they are most often armed, it is sometimes dangerous to pass by them, such is the state of their religious frenzy.[29] It appears that sometimes others provoke these fanatic devotees. On July 9, 1828, a few days before the beginning of the Moharram of 1244 H., the Bombay Police promulgated, in accordance with the 1827 Regulations of the Government, an Ordinance, in which, among other things, it was stated that if the Muslims accompanying the *ta'zia* procession were found in a state of

drunkenness and if they excited a commotion or ut-
tered words which tended to create disaffection
among people, they would be put into prison. Persons
who molested the Muslim processionists and threw
stones and mud at them would also be arrested;
similar action would be taken against those who disturbed the peaceable procession of the horse, which
was taken out on the last night of the ceremonies.[30]

We have seen already that on the tenth day, replicas
of Husain's coffin are taken out to a specified place
for burial or immersion in a river. In this colourful
procession, they parade horses and even elephants,
but as for the horse mentioned in the Bombay Police
order, it is a stuffed dummy, pierced with arrows all
over and it is supposed to represent Husain's horse.[31]

The water, which is an important part of the feast,
is called *sabil*;[32] it reminds one of the absence of it:
'The most precious when one does not have it and
the least needed when it is in abundance'[33]—the
shortage of which caused Husain the most terrible
suffering at Karbala. Among the dishes which are distributed among the poor there is a special one for this
occasion, called *gainj*.[34]

Jawan's narrative given above is completed by the
faithful report of the Muslim writer Afsos,[35] on what
happens in Calcutta on this occasion:

On the seventh of the month of Moharram, [he
says] the Muslims of Calcutta who wish to participate in the festival of the *Ta'zia* or mourning,
gather with banners and flags and proceed to the
place designed for the assembly, with piercing cries
and lamentations and then return to their respective houses. The streets are so packed with people

that he who gets into the crowd by chance has to surrender himself to the pushing and jostling and is unable to go where he had wished to. The crowd inundates the city from noon till night, celebrating with piercing clamour the tragic end of the Prophet's grandson. In Calcutta, people call it the Mourning of the Noon, *dupaharia matam*. On this solemn day Muslims, both men and women, carry to the larger Imambaras as well as to the smaller ones, the offerings of roasted fowl, bread or fried rice and over these offerings get the *fatiha* for Husain recited. On this day so many fowls are slaughtered that in the streets streams of blood can be seen running.

On this occasion low-born Muslims indulge in ridiculous practices in order to get a wish granted. Some of them come to the Imambara with stoves on their heads with rice-pudding cooking on them. Others come to the hall with something like locks on their mouths which resemble small brooches or a horse's bit. These locks are supported by two plates of steel, which are pushed into the jaws, often ripping them. Insensitive to the pain, these stupid persons go round and round the cenotaph and if the lock detaches itself on the third or the seventh turn, they conclude that God has acceded to their wishes and these petty fellows cry out in their excitement and say it is a miracle. The man with the rice-pudding cooking over his head wishes people to believe that he is shivering with cold and he covers himself even if it is excruciatingly hot outside. What is most funny is the belief of these superstitious people that if they indulged in their buffoonery in an Imambara other than the one they had wished to go to, the rice would not boil and the lock would refuse to fall. It is not necessary to believe that educated people can prevent these foolish practices. Even if the Imam, whose martyrdom they celebrate so foolishly, appeared among us he might not succeed with them. Someone has rightly said, everybody has his grain of folly. Thus pass the first ten days of Moharram.

What should strike one in the preceding lines is the imitation of Hindu paganism, which the Muslims yield to in the Imambara. Sealing the mouth with locks is a practice common among Hindu *faqirs*, pictures of which can be seen in books dealing with India.

I am not aware of any special *fatiha* for Husain but in the Muslim prayer book printed at Calcutta, there is only one for both the brothers, Hasan and Husain.

Let the Eternal be pleased to accept the wish I make for the peace of the glorious souls of the two brave Imams, the two beloved martyrs of God, the innocent victims of malice, the blessed saints Abu Muhammadul Hasan and Abu Abdullahul Husain and the fourteen pure persons[36] and the seventy-two martyrs of the plains of Karbala.[37]

The Month of Safar
Ceremonies in commemoration of Muhammad's convalescence

In the month of Safar, Muhammad, the friend of God, fell ill. The vehemence of pain continued for thirteen days and it could recede only at the intervention of the Creator's bounty. It is for this reason that the Muslims consider inauspicious the first thirteen days of this moon. On the thirteenth, which they call violence, *taizi*, they have the practice of making in the Prophet's name offerings of chick-peas and wheat, which they distribute in small quantities.[38]

The last Wednesday

The last Wednesday is considered by the Shi'as inauspicious; the Sunnis, on the other hand, rejoice on this day. While the former dare not move out of their houses, the latter go out for strolls in gardens, where fairs are held and where various spectacles charm the sight.[39]

In the Muslim prayer book printed in Calcutta under the title, *Hidayatul Islam*, or *Guide to Islamism*,[40] there are two special prayers for this day, which are as follows:

We address you with respect and submission. You know what is there in their hearts. We seek your mercy, O the most Generous of generous beings.

The second prayer is repeated while the water meant for offering is drunk: it is a practice to drink a little of the water used for ablution after having purified oneself.[41] This prayer consists of the following verses of the Quran:

Salutations to you. Enter Paradise joyfully to live there eternally.[42] Such is the greeting of the merciful Lord.[43] Salutation among the created beings to Abraham, to Moses and Aaron, to Elijah;[44] peace till the dawning of the day.[45]

'The Persians,' [says Chardin],[46] 'call this Wednesday, *char shamba suri*, the Wednesday of the Trumpet, in other words, the Wednesday of the end of the World, the day when the four great angels Gabriel, Michael, Raphael and Israel will blow the trumpet in order to raise the dead. They consider this day unlucky, that is why they do not transact any business deals, and do not even leave their houses on this day, fearing that whatever they might start will end up sadly. Because of that they hold all Wednesdays unlucky. Caravans do not set out and even shops remain closed on Wednesdays.

Moreover, the 28th and 29th of this month—whatever day of the week they may fall on—are considered unlucky; the 28th, because it was on this date that Hasan was poisoned by his wife, and the 29th, because harmful vegetables are supposed to grow on this date.[47]

The Month of Rabi'ul Auwwal

Ceremonies pertaining to the Prophet's death On the 12th of Rabi'ul auwwal the death anniversary of the Prophet is observed in India. What is however remarkable is that the Turks celebrate on this day the *maulud*, or the birth of the Prophet.[48]

The Moon of Rabi'ul auwwal [tells Jawan] 'is also called Baharia and by the vulgar, The Twelve Dead. According to the Sunnis, God's friend (Muhammad) left this perishable world on the twelfth day[49] of this month. As it spread into the world, this stunning news produced a general consternation and everybody rushed to offer to God their vows and prayers for the peace of the Prophet's soul. This

holy exercise continued for 12 days and it is undoubtedly for this reason that Twelve Dead is the
name given to this month. It is probably in imitation of first Muslims that Indian Muslims assemble
to repeat the act mentioned above.

However—and it is necessary to point out the difference of opinion in this regard[50]—the Shi'as[51]
believe that the Prophet died on the 28th of Safar,
that is to say, thirteen days earlier.

The name *bahāria,* or pertaining to the Spring, given
to this month in India, is a translation of the Arabic
word *rabi',* which means spring. It is necessary to
recall that the old Arab calendar was solar and it was
divided into six parts, as that of the Indians; thus we
have the *First Spring, rabi'ul auwwal,* and the *Second
Spring, rabi'us sani,* for the two months of Spring,
which have become, in the lunar calendar of the Muslims, the third and fourth months, and which can fall
in any season of the year.

The explanation offered by Jawan for the expression Twelve Dead is not very clear, but I do not have
any other explanation, either. This term is found in
the very common proverb *Bara wafat ki khichri aj hai
to kal* [52] *nahin,* that is, the *khichri*[53] for the Twelve Dead
is for today, and not for tomorrow. This means that
abundance may not continue for ever.[54]

The Month of Rabi'us Sani

Fête of Miranji 'Miranji', a double honorific, is a compound of two words. *Miran* is a Persian plural used
to express respect for a person and is derived from
mir (from amir) an Arabic word which means Prince
and which is conferred on Syeds or the descendants

of Muhammad. *Ji* is Hindustani and corresponds to our monsieur or the English master or esquire. This saint is especially designated as Mohiuddin (Arabic), *one who animates religion;* but this is only a title, which is often followed by the epithet *ghausul azam,* and which in India is given to great contemplatives, who have acquired the power to fall into ecstasy.[55]

The following *fatiha* which is recited for this saint gives us the names of his father, mother, brothers and sister:

Syed and Sultan,[56] *faqir* and *khwaja*,[57] rich and poor, king and *shaikh*,[58] dervish and saint our Lord Mir Mohiuddin, whose father is Syed Saleh Zangui, mother, Bibi[59] Fatima II, sister Bibi Nasiba and brothers Abdul Razzaque and Abdul Wahhab [*sic*]. Let our wishes be granted through his intercession.

With this short prayer, the faithful shall recite the first *surat* of the Quran once, the 112th *surat* for fifteen times and the prayer *durood*,[60] eleven times.[61]

The famous Hindustani poet Wali has dedicated to this saint a *qasida*,[62] full of oriental figures of speech and allegories from which the following extracts are worth quoting:

The glory of Islam comes from you.... I hope you will always flash on my mind the torch of your spiritual doctrine.... Mohiuddin is your blessed and celebrated name; it is luminous like the sun. The place around your tomb where pilgrims assemble presents the image of paradise; your regard, there, gives birth to the season of spring. The dust of the threshold of the shrine where you are buried is dearer than the *surma* of Isphahan.[63] The *shaikhs* who come here to pray are sure to have access to the presence of God. He who runs his forehead on the marks of your feet will be as resplendent on earth as the day star. Only contemplatives can comprehend the secret meaning of your words, which they consider equal to those of the Prophet, even that of the Quran.[64] Your help gives strength to the weak and riches to the poor. Could men ever have hoped for remedies, had Hippocrates not received his science from you? You have, in the natural order of things, such power that no obstacle can stop you. A simple doubt has for you the power of an axiom, because you share the secrets of God. Compared with you, Plato and Avicenna are babies. Let the Jews and Christians express enviously their great vexation against the one who sings about you; as for me, I shall be blessed in the two worlds, if you accept Wali's *qasida*, however unworthy it may be for you. All those who listen to the verses I dedicate to you, will be charmed in the same way as they are by Anwari and Khaqani.[65]

The following is Jawan's[66] description of the festival of this famous saint:

The month of *Rabi'us sani* is called by the people the *Moon of Miranji* also, because they celebrate on the eleventh of this month the death of the great Muslim saint who has drawn up the rules of conduct for *pirs* and their disciples, and through whom the whole world is filled with celestial favours. On this day, Muslims belonging to the class of *shaikhs* and some Shi'as also gather round his tomb, and submit to God's wishes, both spiritual and temporal. They recite the *fatiha* for this saint on dishes of sweets which they later distribute respectfully among those present. In addition to the contemplatives who are devotees of this saint or who belong to the order whose head he is, there are a number of Muslim musicians and singers who with their talent add to the brightness of this fête. Moreover, young women dancers come to adorn the occasion with the charm of their attractiveness and the grace of their dance.[67] Thus the whole world assembles there on the 11th of *Rabi'us sani,* and presents an enchanting scene to the eyes. On the one side we see preparations for the banquet made of offerings to the saint and on the other the expression of religious respect by the dancing girls. The musicians play on the *dholki* [68] and the *sarangi* [69] and beat time by raising hands. The singers utter these words in cadence: 'Oh! Oh! the poor *pir* persecuted by a king.'[70] Upon these words those participating in the banquet appear to get agitated and distressed. While one quite beyond himself, falls and rolls on the ground like a sacrificed animal, another appears to be afflicted and starts crying without end, groans loudly and exhales cold sighs.[71] One here with his head bowed gives the piercing cries of '*Hou*',[72] while someone who appears to have renounced life, turns to another person and falls at his feet. While this is happening nobody should leave the place, because it is considered necessary that ceremonies for the saint must be allowed to be completed. They leave only when everything is over. This is generally how they celebrate the festival of saints at their tombs.

In spite of all that we have read of the eminent saintliness of Mohiuddin and the devotion that permeates his fête, Roebuck, in his *Oriental Proverbs*, gives a note in which without any authority cited, the saint is shown as a licentious blackguard. More than anything else this note has the air of a tale from *The Thousand and One Nights*. This is what he says:

Miranji, also called Shaikh Saddo, lived in Sambhal, in Rohilkhand, and according to some at Amroha, in the province of Delhi. He pretended to have great skill in making amulets and in the art of fortune-telling, which is called *ilm-i taksir*.[73] Once a peasant found in the field a lamp with four wicks, made by a famous magician of ancient times. It was a magic lamp; every time that it was lighted, there appeared four genies or familiars, ready to carry out the orders of the person who had lighted the lamp, and visible only to him. The peasant gave this wonderful lamp to the Shaikh. When the Shaikh lighted it, he was alarmed to see the genies and tried to extinguish the lamp; but the spirits told him that once called they would not return without taking an order to be executed. Lascivious by nature, the Shaikh ordered them to bring a beautiful damsel, whom he had seen in a distant country; the order was executed without any delay. The woman, who belonged to a noble family, was extremely surprised and was seized with fear when she found herself in an unknown place and in front of a stranger. However, while he was going to satisfy his impure desire by force or through persuasion, one of the genies warned him that they would obey him only so long as his actions were within the limits of virtue and that if he tried to cross these limits of virtue they would put him to death. For a while, he gave up his plan, but the same thing happened several times and in the end the violence of his passion overran fear; he satisfied his lust and was consequently put to death by the genies. Others say that he got several women in the same fashion and enjoyed them, but the last victim, the daughter of the King of Constantinople, knew his name and address, and on being informed by her, the father wrote to the King of Delhi, who got the Shaikh killed. The lamp was filled with earth and thrown into a river.

This blackguard had, however, the reputation of a saint, even that of a prophet, because of the supernatural powers he had acquired through the spirits, and a splendid *dargah* or tomb is built in his memory at Amroha. According to some, he became a powerful spirit or *jin* after his death and would inspire particularly women by endowing them with knowledge of the future and with other supernatural skills. There are spirits of other dead *pirs*, such as Shah Daryai,[74] Zain Khan, etc., but as they are of an order lower than Miranji's they withdraw when he appears. Hence, the proverb,'When the Mir comes the Pir retires',[75] which means that the subordinate should make place for the chief when he arrives.[76]

The Month of Jamadiul Auwwal

Fête of Madar Madar is the most celebrated Muslim saint of India. The Hindus are one with the Muslims in observing his cult, which the more orthodox call *duli*. The enthusiasm he has aroused among them is expressed in the oft-quoted proverb, 'What hurt will Madar feel, if Shuja goes to Ajmer.'[77]

Syed Badi'uddin[78] Qutbul Madar was the son of Syed Ali (of Aleppo), son of Syed Baha'uddin, son of Syed Zahiruddin, son of Syed Ahmad, son of Syed Muhammad, son of Syed Isma'il, son of Imam Ja'far Sadiq, son of Imam Muhammad Bakar [*sic*] (Baqir) son of Zainul'abidin, son of Imam Husain, son of the Prince of the Faithful, Ali.

Born in Aleppo in 442 (1050–51), he made the pilgrimage to Mecca and Medina at the age of 100 and received from the (Prophet) Muhammad permission to hold his breath.[79] During the reign of Sultan Ibrahim Sharqi, Muhammad ordered him to go to the village of Makanpur,[80] which had become a wilderness on account of the devastations wrought by an evil spirit, Makan Deo. Madar went there, re-imprisoned the genie,[81] made the place habitable and named it Makanpur, or the city of Makan, a name which continues till today. This prophet[82] passed his time there in religious exercises. He had also the power to work miracles, which was known all over India, and which brought there people from all parts of the country. He had fourteen hundred and forty-two sons, three of whom were born of the same mother. He died on the 7th of Jamadiul auwwal, 837 (20th December, 1433) and because of his reputation as a pious man and his power of working miracles, the anniversary of his death has been celebrated since then with a large number of people gathering at Makanpur. This prophet lived for three hundred and ninety-five years and twenty-six days. His tomb was built by Sultan Ibrahim Sharqi.

The above note is provided by Karimuddin, a Madariah *faqir;* madariya, that is to say, a *faqir* belonging to the order of Madar. Lord Valentia included this note in his *Voyages,* vol. I, p. 477, but it does not find place in the French edition of the book. This note appears to be essentially correct, if, of course, one allows for the enthusiasm that has moved the writer's pen. In its description of the genealogy and location of the birthplace of saints the above note is corroborated by the *fatiha* for the saint which is recited at the tomb in the following words:

> By the pure soul of the pivot[83] of the contemplatives and the spiritualists, the shelter of celestial joy and light, the centre of blessed *pirs*, in other words, Pir Badi'uddin Zindah Shah Madar (May God sanctify his grave), by the pure soul of his father Ali Halabi[84] and his mother Bibi Khasul Muluk,[85] known by the name of Bibi Hazira, (I seek from God this Grace.)
>
> The faithful will recite with this intention the first chapter of the Quran, once; the one hundred and twelfth, three times, and the prayer, *durood,* three times.

The 1402 sons, are of course, the saint's spiritual sons or disciples; and this is not difficult to understand. As for the length of his life, which, according to his biographer, is four centuries, it is related to the belief in holding the breath and to the fact that the time of his birth being not known, one feels pleased to distance it from the time of death, which alone is certain; one likes to find among saints the perfection which is denied to ordinary mortals. A little later, there will be reference to another saint who is said to have lived for more than three hundred years.

I should now turn to a description of the festival held in honour of Madar; this is how Jawan[86] describes it:

Common people, especially women, ordinarily call the Moon of Jamadiul auwwal, Madar. However, Madar is the nickname of the saint known by distinguished people under the honorific title of Badi'uddin, but much more widely known as Madar. For this solemnity also they use pikes. Those who wish to participate in the festival, plant them in their respective towns. Musicians beat a kind of big drum, *dhol*, while *faqirs* dance, walk through fires lit up for this purpose, singing songs in praise of the saint.

Madar's tomb is situated in Makanpur, where devotees from far off places converge on the 17th of Jamadiul auwwal, the date fixed for his fête. A huge crowd fills the town; pikes are raised on all sides, and in the night a great number of lamps and lanterns dispel the darkness. Then they all carry the pikes to Madar's tomb and each one seeks grace or fulfilment of a wish.

In this quotation we find another example of the adoption of Indian ceremonies and practices in Muslim cults. The practice of running through fire is evidently borrowed from the Hindus, among whom also there is a fête in which the main rite is crossing this element, which is deified as Agni, *agan*, and this rite is called *dhamal*.[87]

Madar's tomb is raised in the middle of a large square building at each side of which there is a window, which is opened occasionally. It is like any other tomb and is covered with golden cloth. Over it there is a canopy of the same fabric, which is heavily perfumed with the essence of rose, *itr*.[88]

It is said that over the tomb there is a rock, which is suspended by means which are not known. From this comes the proverb: 'There is a row of bricks, but it needs the breath of Madar,'[89] which applies to those who attempt to do the extraordinary, regardless of their capacity.

Afsos gives us greater details than does Jawan:

It is in Makanpur, a village in Kanauj *sarkar*, that we find the tomb of Syed Badi'uddin, known as Shah Madar. He is venerated very much particularly by low-born people, because the *faqirs* who belong to his order come from this class, considering that they are completely illiterate. Moreover, the *faqirs* called Independents,[90] assert that the former's claim to spiritual ancestry is not established beyond doubt. Whatever the truth, these stupid devotees of this saint have chosen black as their distinctive colour. Carrying black flags on gilt pikes, they pass through cities, making a loud noise. This noisy procession is taken out especially in the month of Jamadiu auwwal. Every year in this month, a large number of men and women, generally from the low classes come from far-off places to the village of Makanpur. The pilgrims, led by *faqirs* belonging to Madar's religious order,[91] march in line, most of them carrying standards similar to the ones we have just described. Some of them play the instrument called *rabab*.[92] This procession is called *chhari*,[93] which indicates that they carry pikes. They also give to it the general name, *maidni*.[94] The pilgrims stay near the saint's tomb for several days, busy offering the saint presents and asking for boons; at the end of the seventeenth day of this month, they return home.

The practice of visiting Makanpur is quite old but it is not known who established it. However, it can be presumed that it was started by illiterate and low-born people, as is evident from the hateful crowd which goes there and which imagines that this pilgrimage is more important than that of Mecca. In spite of all that has been said, one cannot form any reasoned opinion regarding the genuineness of Madar's saintliness. In his book *Majalisul Muminin*,[95] Qazi Nurullah Sosatri places him among Shi'as or Imamis,[96] but God alone knows the truth.[97]

As we have seen in the preceding note, Madar is the patron of the order of *faqirs*, who call themselves Madaria, or belonging to the sect of Madar, *madar panthi*. These dervishes have several things in common with Hindu *sanyasis*. Like them they go about naked through all the seasons, and plait their hair, rub their bodies with cow-dung ash and carry iron chains around their waist and neck.[98] The scholar, H. H. Wilson, assures us that they are Sunnis;[99] the black colour that they choose for their flags is the effective proof, because black is the colour of the Sunnis, while green is the colour of the Imamis, or Shi'as.[100] However, since Madar was a descendant of Husain, it appears to prove that he was a Shi'a, and as a matter of fact,

Afsos tells us that he was considered a Shi'a, according to a book he refers to. According to Mr Wilson, the principal practice of the Madariahs is the use of *bhang* (intoxicating liquor extracted from hemp leaves or the exudation of its flowers), in the hope of conjuring visions. According to him, they believe in the divine mission of Muhammad, but they do not have great veneration for his title of prophet and show little respect for the institutions created by him. According to their legends, Muhammad could have access to paradise only by virtue of the words, *dam madar*, breath of Madar, which is the watchword of the sect and to which tradition attributes many miracles. The words *Dam Madar* are also a sort of war cry used by Muslim soldiers.[101]

The Month of Jamadius Sani

Fête for Mo'inuddin Chishti He is one of the most famous of Indian Muslim saints and his tomb is even today constantly surrounded by a crowd of pilgrims,including Hindus.Fanatics take away stones and bricks from his mausoleum and put them in their houses, which in turn become places of pilgrimage on account of these relics.Mahaji and Daulat Rao Scindia, although Hindus and strict followers of the Brahmanic cult,make rich presents at the saint's tomb and to the Muslim priests or the *pir-zadas*[102] attached to the tomb.[103]

The month of Jamadius sani [says Jawan][104] is called by the uneducated people, Khwaja Moi'nuddin, after the greatly famous Muslim saint, Khwaja Moi'nuddin Chishti, who died in this month. This distinguished Pir's tomb is at Ajmer.

It is here, I have been told, where the bows of emperors are tightened by themselves. The ceremony of pikes is observed for this saint also; everywhere, everybody is eager to hoist them. They make it a duty to go for a pilgrimage to Ajmer in this month and if they are unable to make it, they, at least, raise the pikes (at home).

Let us listen to Afsos:[105]

Khwaja Mo'inuddin Chishti, the essence of the con-templatives, was the son of Ghiasuddin Chishti and of the line of Husain and thus a Syed. He was born in Sajestan in 537 H.(1142–3.) When he was 15 he lost his father, but the mystic Ibrahim Qanduzi took a liking to him and made him familiar with the spiritual doctrines and he decided to follow the path of contemplation. He was not slow in plung-ing himself into the most fervent devotion and the most trying practices of austerity. At 20 he gained a few things in religion in the company of Shaikh Abdulqadir Gilani.[106] Later, when Sultan Shahabud-din Ghori conquered Hindustan, and came to Delhi, Mo'inuddin retired to Ajmer to live in retreat. He was followed by a large number of people, who by following his advice succeeded in their spiritual quest. He died there on Saturday, Rajab 6, 636 (12 February 1239) at the age of 97. His tomb exists even today in this town, at the bank of the Jahlara, and it attracts a very large number of pilgrims. No king who reigned in India after his death ever missed placing offerings at the sacred tomb. One may mention, in particular, Jalaluddin Muhammad Akbar,[107] a King extremely religious, who went several times on foot from Agra to Ajmer in order to pay a visit to his tomb and to the tomb of Husain Mashhadi, nicknamed Khing Sawar.[108] The latter was probably a Shi'a, so was probably, Mo'inuddin, a fact suggested by some of his verses in which he expresses his love for the saint, Amir Ali.

The reason for Akbar's pilgrimage has not been indicated by Afsos but we find it in the Memoirs of Jahangir (Sultan Salim):

Till the age of 28, my father, as he himself said, did not have a child who could survive the astronomical hour, and this was a matter of great sorrow to him. Thus he offered at the throne of the Almighty many earnest supplications in order to obtain the object of his desire. As he was still languishing in this state of anxiety, one of his Amirs who had an unbounded respect for dervishes, and the confidence which he reposed in the influence that people of this class wielded, told him one day that there was a *pir* or recluse saint, who lived near the tomb of the respected Mo'inuddin Chishti of Ajmer, and who was distinguished for his purity of life and of conduct, in which, said the Amir, he had no equal, not only in India but in the whole world. In the warmth of his zeal and hope, my father declared that if Providence gave him a child who could survive, he would go on foot from the Capital, Agra, to Ajmer, a distance of not less than 140 *kos*, only with a view to carrying out his wishes, and making his offerings at the tomb of the saint. As my father's wish came from a sincere heart, precisely six months after the death of my infant brother, on Friday, Rabi'ul auwwal, 17 of 978 H.

(18th August 1570) the Very Highest made the writer of this *Memoir* enter the scene of existence.

True to his promise, my father, who now inhabits the celestial world, set out from Agra, accompanied by some of the most notable of the Amirs of the Court and walking 5 *kos* a day and on his arriving at Ajmer, he presented himself at the tomb which encloses the remains of Moi'nuddin. After he had completed his prayers, he immediately decided to go and find the dervish, whose piety and merit were like a debt from which one could get out only through ardent suppplications. The pious recluse was called Shaikh Salim, and my father having reached there, put me in his arms entreating him to pray to God for the survival of the dear child... 'Since you have put the infant in my arms,'said the dervish, 'I name him Muhammad Salim.' Accepting this expression of interest on the part of the dervish as a happy augur, very favourable to his hopes, my father returned to the Capital, from where he continued to have, for fourteen years, a correspondence and a very intimate relationship with the recluse saint.[109]

Sikri, the village where Shah Salim lived, has become since then a town called Fatehpur Sikri.[110] One can even now see the saint's tomb at the centre of a raised hill surrounded by a majestic arcade.[111]

Chishti is the patronymic surname of the line of saints whose head is Mo'inuddin. Salim Chishti belongs to the same line, as many others who are renowned for their sainthood; among them being the distinguished Syed Shah Zahoor on whom we have appended a note in the second part of this book.

Just like Akbar, the famous Haidar called his second son Tipu Sultan,[112] after the venerated *pir* of Karnatic, for whom he had a special devotion.[113]

The Month of Rajab
Supererogatory fast

Muslims have a great regard for the excellence of the month of Rajab. Those who observe the three-month fast, begin it at the appearance of the Moon of this month. The fast that has been observed for one thousand times is called Hazari[114] in the Muslim world. Since it is considered an act of great merit, most of the people observe it. During this period most of the faithful distribute, every Friday, on wood plates, rice prepared in several ways. The offerings are made in the name of Syed Jalal Bokhari, a saint celebrated up to the horizon.[115]

Ancient Arabs also considered this month sacred and consecrated it to fasting. Waging war during this month was prohibited in the same way as in the months of Moharram, Ziqada and Zilhijja.[116] The three months of supererogatory fasting are Rajab, Sha'ban and Ramazan. The fast continues up to the seventh of Shauwwal, the last seven days being called blank. However, they do not fast on the first Shauwwal, the day of Id Fitr, about which we shall talk below.

The Muslims of India fast also on the 10th of Moharram, which we have described above and also on the 10th of Zilhijja which is the fête of Id-ul azha or Id-Qurban, notwithstanding what M. d'Ohsson[117] has to say on fasting being prohibited on this day.

Syed Jalal, or to be more exact, Jalaluddin Bokhari, who has also the title of shaikh, was the son of Mahmud and grandson of Jalal. He has been called the Lord of the Created Beings, *makhdum jahaniyan*. Afsos[118] informs us that he was born on the night called *barat*, which is consecrated to the memory of the dead,[119] that is to say, Sha'ban 14, 707 H, February 8, 1307. Even though he was his father's disciple, who was himself a saintly person and whose spiritual successor he had become, he received great religious instruction from Ruknuddin Abulfath Suhrawardi.[120] On coming to Delhi, he also profited from the instructions of Shaikh Nasiruddin, nicknamed Lamp of Delhi.[121] He died on Wednesday, Id Qurban, 775 (May 24, 1374).[122] He was buried in Outch, a town in Multan.[123]

The *faqirs* called Jalalia, and Malang,[124] are his disciples or belong to his sect. The latter go about all naked, if one is to believe the author of *Khulasatut Twarikh*.

The Month of Sha'ban
Shab-i Barat or the Fête of the Dead

A grand fête of the Muslims takes place on the 14th of Sha'ban, which is called Shab-i barat, or the night of deliverance. On this solemn day the faithful assemble and make in the name of all those who reside in the Kingdom of Eternity, offerings which are considered inviolable;[125] these offerings comprise bread, halwa[126] and flasks of water.[127]

As they make the offering, they kindle lamps and recite the following *fatiha*, called the *fatiha* of the lamp, *fatiha chiraghan*.

O God, through the merit of the light of the apostolate, Our Lord Muhammad, may the lamp that we burn on this holy night, be for the dead a guarantee of the eternal light which we pray to you for. O God of ours! Deign to admit them in the room of unchanging felicity.

Having expressed the above intention, recite the first and the 102nd *surats* of the Quran.[128]

The ordinary *fatiha* for the dead, which differs from the above, is translated below:

Through the merits of the holy and honest Prophet Mustafa Muhammad (May God be propitious and grant him salvation), through the purity of name and the enlightenment of his mind which redresses sin, and through the purity of the soul of the deceased Named, may God make the day of his mercy and favour shine over his tomb; may He deign to water with the showers of His grace, the earth that covers his body, and grant him heaven as his abode. And may the merit of such an oblation be granted also to those who have some claim to the protection of the Named, to those pure souls among the dead who have enlivened their merit by keeping their faith in God, and to all those who have lived and died in the arms of Islam.

Recite the first and the 102nd *surat* of the Qur'an.[129]

This fête is celebrated in Persia also. This is how the famous Protestant traveller, Chardin[130] describes it:

The Persians say that in this night, through the intercession of Muhammad and Ali, God saves a large number of the souls of their co-religionists, by making the angel Gabriel draw them from hell.[131] They say that there is a great merit in going to pray on this day to the graves of the dead and in distributing alms liberally. This is one of the most popular fêtes. It lasts for three days, and they call it the days of charity and good work. The devotion consists in sitting at the graves and each one talking to the dead parents and friends, calling them with lamentations and prayers; then they smoke and finally eat whatever they have brought, fruits, pastry, sweets, a large part of which is given away to the poor for the benefit of the dead...

The Month of Ramazan
Fasting

By the bounty of the Very High, Ramazan is the month of Muslim fasting. Happy and content, the faithful observe scrupulously this fast with the appearance of the New Moon till the next New Moon is sighted. Men to whom God has given riches as their share never forget to make arrangements for the light feast for the moment of breaking the fast; this feast consists of sherbet made of sugar and rose water, or Egyptian willow,[132] almond, pistachios, stoned dates, and other light milk preparations. The principal meal is taken after this light meal. Then they take rest; but they do not forget to get up in the early hours of the morning and take something light.[133]

Commemoration of Ali's Death

The 21st of this month is the day of the commemoration[134] of the martyrdom of the great saint Ali, who was the successor and brother (cousin) of the Prophet. All those who wish to pay mourning homage to the Elect of God, join an assembly in order to listen to the account[135] of the circumstances which led to this sad event and sing funeral hymns in its memory. Profound meditation presides over the reunion; tears run abundantly from eyes, and cold sighs[136] are exhaled from all hearts.[137]

The Month of Shauwwal

Id Fitr

In the whole Muslim world, the first day of the month of Shauwwal is consecrated to the celebration of Id Fitr, or the breaking of the fast or simply Id. After having offered prayers called *dogana*[138] and full of contentment, the faithful felicitate each other,[139] meet at bright and cheerful gatherings and give themselves up to joy and exhilaration. They visit each other, but one never offers his felicitations to persons of higher dignity without offering some presents[140] and these persons in return give him a dress of honour or make some other gifts. This is how the day passes.[141]

There is no point in giving further details on this fête, which has been described at greater length and with precision in the tableau of the Ottoman Empire by M. d'Ohsson[142] and by several other writers.

The Month of Zi-Qad

There is no fête in this month,[143] that is why it is called empty, *khali*, and is considered inauspicious. Muslims, therefore, do not marry in this month, nor do they contract any other relations.[144]

The Month of Zil Hijja
Id Qurban

Having put on the Ihram, or the robe of penitence,[145] the Muslims make a religious tour of the Kaba, in the month of Zil Hijja, or the month of pilgrimage. Those who do not have the good fortune of performing this ceremony themselves, should at least participate in the fête called Id Qurban (fête of the sacrifice), which is celebrated on the 10th of this month,[146] with the sacred sacrifice of a victim. This grand fête is distinguished by its great joy and open gaiety. Nothing is more agreeable to God.[147]

It would be quite superfluous to give a description of this fête, which is common to Muslims all over the world. M. d'Ohsson and other writers have described it in detail. However, it is necessary to mention that in India they choose a particular place[148] near the mosque, where they celebrate Id. This is a kind of roofless chapel, with small minarets and an altar. This place, which should not be confused with an Imambara, discussed above in the context of Moharram, is called, Qurban-gah, place of sacrifice, or Id-gah, place of Id, where they celebrate this solemn occasion called an antonomasia, Id, fête.[149]

Id Ghadir

In the month of Zil Hijja, there is another great fête, but it is celebrated only by the Shi'as. It is called Id Ghadir[150], august solemnity, which the spirit will forever remember willingly, which the ear will always be pleased to hear the pleasant mention of.

The world unanimously sings praises of the excellence of this fête established to commemorate the unambiguous declaration, which on God's order Muhammad made on this day, according to which Ali, the Amir of the Believers, the King of Saint hood should be his successor.[151] This fête is called Ghadir, on account of the name of the place, Ghadir Khum,[152] where this event took place. Whoever rejoices on this day, shall have the merit of putting his steps in the kingdom of eternity.[153]

Solar Fêtes

The Month of Jaith (May–June)
Fête of Salar Mas'ud Ghazi

The tombs of Rajab Salar and Salar Mas'ud—the latter nicknamed, Ghazi, that is, warrior—are at Bahra'ich.[154] It is said that Rajab Salar was brother of the Pathan Sultan Taglic Shah [*sic*][155] but about Salar Mas'ud there are conflicting views. Some people say that he was a Syed, or a descendant of Muhammad, through Husain, and at the same time a close relation of Sultan Mahmud of Ghazni; while others say that he was a Pathan (that is, an Afghan). Whatever it may be, he suffered as a martyr and his tomb is a place where an immensely large number of people are attracted by devotion. Once a year specially, pilgrims come in a body from far away places. Some of them, ordinarily merchants of low status, come out of their towns and villages with lances decked with flags and led by drumbeaters singing and playing loudly on their instruments. The devotees of the saint take care to reach the tomb two or three days before the first Sunday of the month of Jaith (May–June), which is the day of his fête, or to be more exact, his wedding. It is said that the day of his wedding was also the day of his martyrdom. He was in bridegroom's dress when he was struck. Persuaded by the belief that Mas'ud Ghazi revived his marriage every year, a person from Rudoli, oil-man by caste, would bring annually, to the saint's tomb a litter, a chair and other things of domestic use. This custom has been followed by members of his family for a long time.

The low-born have great faith in him, because they believe that he would place them under God's protection, even if they have not been free from villainy.

Around the shrine in which Mas'ud Ghazi is buried, fanatics tie themselves to trees and hang by their feet, hands and neck with the hope that through this vain act of penitence, they will get their wishes fulfilled. People, always from among the vulgar, call this great person Gajna Doulha,[156] and the women devotees, Salar Chhinala.[157] The reason for these appellations is that the woman who enters the shrine falls down faltering and imagines that the saint has sucked her. Cursed be such a thought! Anathema for this delusion ! The truth is that the upper part of the chapel is lit by a big luminary, the chapel is very small and the entrance narrow and there is the ceaseless pushing and jostling of people who try to enter and get out of it, and there is suffocating heat inside, so much so that those who enter are drenched in sweat. Women being more delicate than men, feel weak and swoon very easily. Whoever says more than this is a liar and impostor. However, it is certain that had

Madar[158] and Salar not come to this earth, the ordinary people who spend here all that they have earned could have otherwise amassed a lot. There would not have been a vegetable-seller or a butcher who would not have become rich.[159]

The above lines make it clear that the reputed saint, Salar Mas'ud Ghazi, also called Ghazi Mian,[160] was a close relation, a nephew[161] of Sultan Mahmud and is one of the two persons, who have the special title of *salar* (a Persian word meaning chief, captain) but he alone is the reputed saint. As a matter of fact, two different words are used to distinguish the tombs of the two persons, *turbat* for the first and *dargah* for the second. The latter refers to a saint's tomb, as we have seen in the 'Preliminary Observations', while the former refers to the tomb of a person who is not venerated by the people, and this proves that Rajab Salar was not considered as a saint.

Afsos has given us the most acceptable account of many things regarding Salar—the reason why the fête is called wedding, the description of the Hindu penance which the devotees practice near the shrine and the satisfactory explanation of the incidents that take place inside the shrine. The justly celebrated scholar Mr H. H. Wilson says that this ceremony is called *ghazi mian ki shadi, the marriage of the Ghazi*, and he thinks that the world *shadi* is a corruption of the word *shahadat*, martyrdom.[162] I would not accept this conjecture. In the first place, I have not come across the word *shadi* in the Hindustani works which furnish the material for this *Memoir*, but only the synonyms *beyah* and *urs*, words which have no connexion with *shahadat*. Secondly, there is nothing absurd in this legend, and so it is not necessary to take recourse to conjecture in order to explain it.

The following extracts provide details of the fête consecrated to Salar Mas'ud, the most celebrated Indian Muslim saint, after Madar, whose fête has already been described.

In the solar month of Jaith[163], a very large number of Muslims hoist banners, which they call lances of the saint, *pir ka naiza,* that is, of Salar Ma'sud Ghazi. The low-born people among Muslims are devoted to this elect of God, whom they have chosen as their patron. Full of faith in him, they repeat his name in the form of ejaculatory prayer or utter these words, 'O Great Saint!' The tomb of this saint is situated in Bahra'ich in the kingdom of Oudh. The brave Nawab Asafuddaulah[164], the priceless pearl of vizarat, does not miss a visit to the place at the time of pilgrimage, where the famous fair[165] is also held.

The day before the saint's fête which is also called wedding, and which is considered a day to be devoted to pleasure, all sorts of banners are planted on the river bank, and under each of them they place lamps fed with clarified butter. Some people fix pikes on their belts and appear so much beyond themselves that they excite the astonishment of spectators. While one folds his hands respectfully, the other jumps for joy; one sighs while the other prostrates himself in prayer. One beholds a thousand acts and a thousand postures. Many people come to ask God's grace through the saint's intercession. While offering flowers and sweets, they say, 'Let my wishes be fulfilled.' All the while, musicians beat cymbals and sing these words: 'He who hears the song in celebration of Gajna Doulha will have his desire fulfilled.'[166]

Every year pikes with green and red[167] flags are thus hoisted over an expanse of several *kos*[168] of ground which is protected from the burning sun by elegant awnings of different colours.[169] They set up two rows of shops, where one can get anything one desires. Young Indian women with the body of fairies and faces like the moon drive around in *manjholi* and *rath gari*,[170] as numerous curious people who have no other aim but to divert themselves, fill the avenues. The saint's devotees, as many of them as can, come to Bahra'ich, at the time we have just mentioned. People go to the tomb on the morning of the first Sunday of Jaith, having made all the arrangements for the fête, a day earlier. What is really remarkable is that the low-born people offer all that is necessary for a real wedding, in the belief that Salar Mas'ud re-enacts his wedding on this day. On the other hand, the guardians of the saint's tomb soak in water[171] the *langui* (*lungi*)[172] which the saint used to wear and which is placed on a couch, in the morning. The water, according to them becomes holier than the water of life[173] and they distribute it as relic and in return get gold and silver. I have not visited the place myself, but I have heard all this described to me a number of times. A hundred persons remain there, tied voluntarily to the tree through devotion. A thousand people, lame, blind, armless, leprous, live there in the hope of being cured. If a hurricane,[174] quite frequent at this time of the year, passes through the town on the day of the fête the devotees do not forget to point out that it is the saint, who as an expression of his benevolence, makes a *dev*[175] sweep the land.

The Month of Bhadoun

Fête of the Bera of Khwaja Khizr The people of the Orient have several views on the origin of Khwaja Khizr.[176] Many think he is Phinées, grandson of Aaron,[177] while others say he is the prophet Elijah;[178] the Turks have mixed him up with Saint George. As a compromise to these three views, some make us believe that he is one soul with three persons. However it may be, according to the Muslims, Khizr is the discoverer of the source of the water of life and its guardian.

Moreover, Indian Muslims believe that he was an expert in the art of divination. From this they would call *khabr-i khizri*, news from Khizr, a piece of news on which one speculates in the same way as the public does about the intentions of the government. In one of his *ghazals*, Wali has mentioned Khizr in this context: 'The news has reached through Khizr that his letter is the ruby of your humid lips.'[179] Khizr is considered the patron saint of the waters and Jawan[180] describes in the following words the fête held in his honour:

In the month of Bhadoun,[181] which has 31 days, all those whose wishes have been fulfilled join together in launching into water a boat in honour of Khwaja Khizr and putting into it, according to their means, offerings, especially of milk and crushed grain. With a lot of ceremonial and reverence devotees of Khizr carry the *bera* to the river bank on Friday evenings and at some places on the Thursdays[182] of this month. After having lighted lamps and candles, people, great and small, make the respectful oblations, while swimmers together push the boat to the middle of the river and this presents a ravishing sight.

In the following ridiculous verse, Wali alludes to the water of life, whose guardian Khizr is and to the *bera*, which is dedicated to him: 'Who could without a *bera* approach this mouth, from which spills the precious water of life.'[183]

The *beras* they launch in rivers in honour of Khwaja Khizr are of two kinds; the bigger ones with the generic name *nao*[184] (*navis, nef*) which are floated annually with great pomp during the fête of Khizr and the small ones which every Muslim puts in the river dutifully with flowers, one or several lamps, etc., in it on Fridays in the month of Bhadoun. Viewed from a distance they are a charming sight.[185] These small *beras* are made of clay[186] and during this season one sees thousands of them on the Indian rivers. Hodges, who was a traveller and also an artist, did not know what it was about and so he was astonished to see this scene. He writes:

While passing through Murshidabad,[187] an evening of a holiday of the Muslims I was very much pleased to see the river covered with innumerable lights which floated on the surface of the river. It was truly an extraordinary spectacle and it was very difficult, at that time, to find out any explanation for it. But I discovered soon after on enquiry that on this occasion Muslims make a large number of small lamps and after illuminating them, float these in the river and as they stay lighted for several hours, the current of the river carries them to a considerable distance.[188]

Travellers tell us that the inhabitants of the Maldive Islands, who profess Islam, launch annually a small boat filled with perfumes, gums and sweet-smelling flowers as offerings to the King of the Sea and leave it to the care of the wind and waves. Nobody would doubt that the King of the Sea in this case is Khizr.[189]

The following are the words of Khizr's *fatiha*:

For physical and spiritual health, I take shelter in the benediction of one who fulfills the wishes of mortals and drives away from them sorrows, that is to say, Khwaja Khizr, the illustrious Elie.

After having expressed his intention thus, the faithful would recite the verset, *fatiha*.[190]

Festival of Goga

Muslims are greatly devoted to Goga whom they also call Zahir Pir. They adore him with their heart and soul and express their humility in various ways. During the month of Bhadoun, they celebrate his fête, marching in the streets with pikes in hand, playing on different musical instruments, singing in chorus the praises of the saint. These processions last for one month. Finally, they assemble at a particular place and all of them plant there their pikes. They hold on this day a large fair, which is remarkable for all kinds of diversions and curious spectacles. I have heard it said that the tomb of this saint is in the Duab, but the practices I have mentioned above are followed everywhere.[191]

NOTES

1. *Bara Masa*, p. 96. We have seen in the *Preface* that Indian Muslims give the name of a principal *pir* to the lunar month in which the fête in his honour is celebrated. It is thus natural that this month of the greatest solemnity is named in honour of one whom Imamis consider the spiritual and temporal king and powerful protector of Muslims. *Wahi din o dunya ka hai badsah: karaiga bakhubi hamara nibah.*
(*Araish-i Mahfil*, p. 2).

2. *Bara Masa*, p. 94.

3. Ibid.

4. The Imamis have such a great horror of this general, who under orders from Caliph Yezid, got Husain atrociously killed, that the name *shimr*, has become for them a word of abuse, synonymous with vile, loathsome, wicked, etc. (Shakespear, *Dictionary*, p. 550).

5. Karbala is a place, dry and arid, situated in Iraqi Arabia, where one can still see, in spite of Caliph Motawakkil's efforts, the tomb or confession, *mashhad* of Husain. Muslims go there for pilgrimage with devotion and have such a great veneration for the earth of that place that they make rosaries of it, which they call *kantha* and which they use with great respect. The big bead is called, Imam, after the priest who leads the prayer, that is, *paish namaz*. Shakespear, *Dictionary*, pp. 68 and 646.

6. In the text we have only the word *abid*, but this refers undoubtedly to the second son of Husain, also called Zainul 'abidin, the ornament of the devout, who being sick was spared and thus survived this disastrous day.

7. These cenotaphs are called *ta'zia*, literally, mourning and it is for this reason that this word is often used for the fête of Moharram.

8. *Rauza khwan*, literally, the reader of the garden (tomb). By this expression one means the person who is employed to recite the praises of Husain. See the *Preliminary Observations*, above.

9. *Salaam*, peace (on Husain).

10. *Marsiya*. This is probably the hymn called *bhatiyal*, which is a sort of elegy for Husain and Hasan. Shakespear, *Dictionary*, p. 148.

11. According to Tavernier, this fête is established in memory of both Hasan and Husain, but he refers to only one cenotaph, and, whatever he says, suggests that he means only Husain. *Voyages*, vol. I, p. 427: Paris, 1677, in-4.

12. *Voyages*, by Chardin, Langlès ed., vol. IX, p. 49.

13. *Shuda* is a word, I think derived from the Arabic *shud*, to attach, which refers to the piece of cloth they attach to the pike, and not from *shud* in the sense of running, and, least of all from *shuhada*, martyrs, which some Orientalists would have us believe.

14. Shakespear, *Dictionary*, p. 251.

15. *Asiatic Journal*, vol. XXVII, p. 102.

16. *Araish-i Mahfil*, p. 111.

17. Shakespear, *Dictionary*, p. 397, 422.

18. Valentia, *Travels*, vol. 1, p. 473.

19. Taken here in the sense of *Khalifa*, the spiritual and temporal head of Muslims.

20. Gilchrist, *Hindoostanee Philology*, p. 307.

21. *Asiatic Journal*, XXVII, p. 355.

22. It is probably what they carry in the procession, because Valentia refers to ornate coffins with rich gilding, which are placed in the Imambaras and which are carried in the procession on the morning of the tenth day. Valentia, *Travels*, 1, p. 473.

23. *Araish-i Mahfil*, p. 130.

24. This Imambara is permanently illuminated with a great number of candles; the tomb is strewn with flowers and the priests recite verses from the Quran day and night. Hamilton, *East-India Gazetteer*, vol. 11, p. 131.

25. *Araish-i Mahfil*, p. 104.

26. Ibid.

27. Muhammad's descendants who wear green clothes do not abandon them on this occasion. Valentia, *Travels*, vol. 1, p. 227.

28. Valentia, *Travels*, vol. 1, p. 473.

29. For example, on the occasion of the festival of Moharram in 1244 (July 1828), the city of Lucknow was the theatre of many unhappy events. For this see the account in *Asiatic Journal*, vol. XXVII, p. 355.

30. *Asiatic Journal*, vol. XXVII, p. 102.

31. Valentia, *Travels*, vol. I, p. 227.

32. Shakespear, *Dictionary*, p. 491.

33. Ali's words. *Muntakhabat-i Hindi*, first ed., vol. I, p. 21.

34. For this word see the *Hindoustani Dictionary* by M. Shakespear, p. 700.

35. *Araish-i Mahfil*, pp. 130, 131.

36. That is to say, Muhammad, Fatima and the twelve Imams. See *Voyages* by Chardin, Langlès ed., vol. IX, p. 487.

37. *Hidayatul-Islam*, p. 268.

38. *Bara Masa*, p. 104.

39. Ibid.

40. *Hidayatul-Islam*, p. 276.

41. See my *Doctrine et devoirs de la religion musulmane*, p. 148.

42. Quran, XXXIX, 72.

43. Quran, XXXVI, 58.

44. Quran, XXXVII, 120, 130.

45. Quran, XCVII, 5.

46. *Voyages*, vol. IX, p. 90.

47. Shakespear, *Dictionary*, p. 559.

48. *Mouradgea d'Ohsson, Tableau de l'empire ottoman*, vol. II, p. 358, ed. in-8.

49. That is to say, 12th Rabi'ul auwwal of the tenth year of the *hijra* (June 8, 632, AD.)

50. Shi'a or Imami. The author of *Bara Masa* belongs to this sect.

51. *Bara Masa*, p. 19.

52. *kal* means yesterday as well as to-morrow in the same way as *parson* means both the day before yesterday and the day

after tomorrow: *tarson* three days after or before, *narson*, four days before or after. The time of the verb indicates the sense of the adverb.

53. See above for what this dish consists of.

54. Roebuck has committed two mistakes in translating this proverb as, *this is the khichhri of the Twelfth of Safar*: first by mistranslating *bara wafat* and secondly by taking Rabi'ul auwwal as Safar. See *Oriental Proverbs*, part II, p. 29.

55. Shaikh Muhammad Ghaus is one of the principal persons who have been given this title. He is buried in Gwalior. According to the author of *Araish-i Mahfil*, he could subjugate even Mars!

56. *Sultan*, which means king, and *Shah,* which has the same sense, are honorifics given to dervishes and *faqirs*. See *Preliminary Observations*.

57. Honorific given particularly to the Mughals. Shakespear, *Hind. Gram.*, p. 142.

58. Honorific which means, literally, old man and is given like Syed to the descendants of Muhammad. Moreover, this title is also given specifically to Muslims of Arab origin.

59. *Bibi*, that is, madame.

60. See *Eucologe Musulman*, p. 222.

61. *Hidayat ul-Islam*, p. 267.

62. *Qasida* is a kind of poetry in which the verses have the same rhyme. See, Gladwin, *Dissertation on the Rhetoric of the Persians*, p. 2. The word, *qasida*, is feminine in Arabic and masculine in Hindustani.

63. *Surma* is the collyrium of lead, the most famous being that of Isphahan.

64. We known that the Muslims regard it as the word of God.

65. Famous Persian poets.

66. *Bara Masa*, p. 24.

67. Afsos (*Araish-i Mahfil*, p. 110), talking of the tomb of Shah Arzan, situated near Azimabad (Patna), says that every Thursday, a large number of people go there, including prostitutes and bayadères, who execute dances there till the middle of the night.

68. *dholki*, a kind of small drum.

69. *sarangi*, a kind of violin.

70. Allusion, perhaps, to what will be said later.

71. *sard*. We should have had hot instead of cold here, but on account of its peculiarity, I have retained this epithet, which often accompanies the word *dam*, sigh, in Persian and Hindustani.

72. *hu*, an Arabic word, which means him (his), and which is also used to address God, as it is the case here.

73. Could be read as *ilm-i taqdir*.

74. Or *daryai*; see in the Second Part below the article on this saint.

75. *Aaé mir bhage pir (Oriental Proverbs*, part II, p. 26).

76. *Oriental Proverbs*, part II, p. 27.

77. *Upri jhant madar ki Shuja chale ajmer* (Roebuck, *Oriental Proverbs*, vol. II, p. 2). Literally, An *pilus partium genitalium* Madari *evulsus erit, si*, etc. The word, *shuja*, literally means brave; but here the word is used in the same sense as in Arabic, where it refers to an unspecified person. Ajmer is the old capital of the province of the same name, where is buried Mo'inuddin, about whom we shall talk in the next article.

78. Badi'uddin. In other words, the marvel of religion and not Badruddin, as has been printed in *Oriental Proverbs* (part II, p. 219), which means the moon of religion.

79. *habs dam*, the practice which *faqirs* indulge in, considering it as a religious act and also as a means of prolonging their life, in the belief that everybody has a given number of breaths, thus the longer the breath, the longer the life. Shakespear, *Dictionary*, p. 365.

80. Village near Firozabad, in the province of Agra. Carey, *Map of Hindoostan*.

81. From that perhaps the proverb *dubli mārain shah madar*, Shah Madar strikes the weak, which refers to the man, who is tyrannical with those who are weak, but does not have the courage to attack persons stronger than him. Roebuck, *Oriental Proverbs*, II, 96.

82. Enthusiasm for Madar alone can explain his being considered such a person.

83. *madar*. The author puns on this word, which is the saint's name and which also means, pivot, home, centre, etc.

84. The text has *chalbi* but I think, the correct word should be *halabi*, that is, from Alep.

85. *khasul muluk*, literally, close to the kings.

86. *Bara Masa*, p. 33.

87. Shakespear, *Dictionary*, p. 427.

88. *Voyages de Valentia*, Fr. trans., vol. I, p. 285.

89. *īnt ki pānt dam madār*. Roebuck, *Oriental Proverbs*, part II, p. 219. See below the explanation of *dam madar*, the breath of Madar.

90. Azad. They shave their beard, eye-brows and lashes, and take a vow of chastity. Shakespear, *Dictionary*, p. 38.

91. *madaria*. Madaris, belonging to the sect of Madar. See below.

92. *robab*, a kind of of violin.

93. *chhari* is the name of the pike carried in the procession of Shah Madar's devotees and in other similar processions. From that, the word has come to mean this procession itself. These pikes are also called *jhanda* and this procession *Madar Jhanda* (*Asiatic Journal*, N. S. vol. IV, p. 75).

94. *maidni*, group of people who go to a saint's tomb.

95. *majalisul momnin*, assemblies of the believers.

96. The author of this note belonged to this sect. See *Preliminary Observations*.

97. Ibid., p. 76.

98. *Asiatic Journal*, N.S. vol. IV, p. 76.

99. Ibid., p. 75.

100. Ibid. De Sacy, *Chrestomathie ar.*, vol. I, p. 49, new ed.

101. *Asiatic Journal*, N.S. vol. IV, p. 75; *Asiatic Researches*, vol. XVI, p. 135.

102. *pir zada*, literally, son of a *pir*.

103. Hamilton, *East-India Gazetteer*, vol. I, p. 28.

104. *Bara Masa*, p. 38.

105. *Araish-i Mahfil*, p. 150.

106. That is to say, the province of Gilan, *gilan*, in Persian, from which is formed the word, *gilani*. In Arabic this province is called *jilan*, from which is formed *jilani*, which is synonymous with the word first mentioned. See in the Second Part of this *Memoir* the article on this saint.

107. The author here refers to Akbar the Great who, so Father Catrou in his *Histoire du Mogol* thinks, was almost a Christian. According to this writer, Akbar did not believe in his own religion, openly protected Christianity, had in his palace the statue of the Holy Virgin, etc. I believe that one must not trust Father Catrou's assertions. The principal object of his writing this book was to give importance to the apostolic work of Jesuit Fathers in the Empire of this Mughal.

108. *khang sawar* that is, riding a grey horse.

109. I have borrowed these lines from the excellent article which the illustrious Orientalist M. de Sacy has written on the English translation of the *Memoirs of Jahangir*, by D. Price.—*Journal des savants*, 1830, p. 362 sqq.

110. Afsos says,'Sikri was a village 12 *kos* from Agra. Akbar built there a castle of stones at the order of Shaikh Salim Chishti and also several beautiful edifices, monasteries, mosques: Then he gave it the name of Fatehpur (the City of Victory), and he made it his capital (that is to say, the place of his residence)' (*Araish-i Mahfil*, p. 74).

111. Hamilton, *East-India Gazetteer*, vol. I, p. 553.

112. Tipu means tiger in Canara, or perhaps, lion; because in Hindustani, they practically always confound these two animals. *Sher* would mean either a tiger or a lion and it is in the second sense that it is used as a proper name, while *singh*, has only one meaning, lion.

113. The tomb of this saint is at Arcot and is a place of pilgrimage. Hamilton, *East-India Gazetteer*, vol. II, p. 271.

114. *hazari* from *hazar*, one thousand.

115. *Bara Masa*, p. 59.

116. Sale, *Observations historiques et critiques sur le Mahometisme*.

117. *Tableau de l'empire ottoman*, vol. III, p. 10, ed. in-8.

118. *Araish-i Mahfil*, p. 166.

119. See the following article.

120. We shall talk of this saintly person in the article on Zakaria.

121. See the article on Nizamuddin.

122. See the article on the month of Zil Hijja.

123. *Araish-i Mahfil*, p. 166.

124. Shakespear, *Dictionary*, p. 209.

125. *achhuta*, that which should not be touched; epithet used for the food offered to the dead.

126. *halwa*, kind of soft pastry, made with flour, clarified butter, *ghee*, and sugar.

127. *Bara Masa*, p. 65.

128. *Hidayat ul-Islam*, p. 272.

129. Ibid.

130. *Voyages de Chardin*, vol. IX, Langlès ed., p. 140.

131. Or to put it better, the purgatory.

132. *baid mishk*. See in *Les Oiseaux et les Fleurs*, the note on p. 142 sqq, where I have explained what is to be understood by *ban*, a synonym of *bed mishk*. I refer to this note with all the more confidence because it has earned the approbation of the principal Orientalists of Europe, particularly of the celebrated, M. de Sacy. See his *Chrest. arabe.*, new ed., vol. I, p. 258.

133. *Bara Masa*, p. 74.

134. We know that the Shi'as or Imamis consider the first three Caliphs as pretenders. The author who belongs to this sect, speaks in conformity with these principles.

135. Perhaps the same as we find in *Gul-i Maghfirat*, p. 47 sqq, and I do not quote it here because its details are given by Herbelot and other writers. The same fête is celebrated in Persia also. See Chardin, vol. I, p. 208.

136. We have already come across this expression.

137. *Bara Masa*, p. 74.

138. *do gana*, prayer in which the body is bent twice.

139. Almost similar to Easter as observed in several Christian countries.

140. In India one does not present oneself before a superior without a present in his hand: however, this practice has been abolished in British India (*Asiatic Journal*, vol. XXVIII, p. 631).

141. *Bara Masa*, p. 79.

142. Under the name, *beyram*, which means fête in Turkish in the same way as *Id* in Arabic.

143. By mistake, Chardin has said that Id Qurban takes place on the 10th of this month, an error, which the late M. Langlès did not take care to point out. See the *Voyages de Chardin*, vol. IX, p. 7 and elsewhere.

144. *Bara Masa*, p. 85.

145. For this cloth see my *Exposition de la foi musulmane*, p. 84.

146. This is the fête which the Turks call *Qurban-beiram*.

147. *Bara Masa*, p. 89.

148. The same as *minhar* in Arabic, which appears to be different from the *musalla*, place in the open air, where people assemble to offer prayers on certain occasions. M. de Sacy, *Chr. Arabe.*, vol. I, p. 192.

149. Shakespear, *Dictionary*, p. 601; Rousseau, *Dictionary*, p. 90; Hamilton, *East-India Gazetteer*, vol. II, p. 723.

150. *id-i ghadir*, or the fête of the pond.

151. *wasi mustafa*, that is, inheritor or representative of Muhammad, one of the titles of Ali. See *Bibliothèque Orientale*, for the word Ali.

152. *ghadir khum*, the place where the caravans halt, halfway between Mecca and Medina, where there are small pits always full of water. *Chr. arab.* by M. le baron de Sacy, vol. I, p. 193. The same fête is celebrated by the Persians. See the *Voyages de Chardin*, Langlès ed. vol. VI, p. 310.

153. *Bara Masa*, p. 89.

154. 'Ancient city in the kingdom of Oudh, situated on the bank of the Sarju. It is extremely vast and important. In its environs one can find so many mango-groves, and beautiful gardens surround it on all sides' (*Araish-i Mahfil*, p. 97).

155. And father of Sultan Firoz, King of Delhi. See *Ain-i Akbari*, vol. II, pp. 33 and 104, ed. in-8.

156. *Gajna dulha*, the joy of the married male: from *gajna*, to be happy and *dulha*, bridegroom.

157. *Salar chhinala*, Salar, the libertine.

158. See above, the article on this saint.

159. *Araish-i Mahfil*, pp. 46, 47.

160. *miyan*, is a title of honour equivalent to monsieur. This is also an expression of amity, with which a husband or a lover is addressed.

161. Shakespear, *Dictionary*, p. 581.

162. See *Asiatic Journal*, vol. IV, N.S., p. 75.

163. *jaith*, the second Indian month, which commences on 9-13 May and ends at the same time in the month of June.

164. King of Oudh, who reigned from 1756 to 1775. He has been celebrated by Sauda and Hasan of Delhi and Mir Taqi of Agra, Hindustani poets, who have enjoyed a very great reputation and whose works have been published in Calcutta. In the Introduction of the translation of *Conseils aux mauvais poetes*, by Mir Taqi, which I have published, I made the mistake of making him the contemporary of Shah Alam I, son of Aurangzeb, when, as a matter of fact, he lived in the times of Shah Alam II, who reigned from 1761 to 1806.

165. We have seen a description of this in 'Preliminary Observations', above.

166. See above the explanation of this name, which the common people have given to this saint.

167. Green is the colour of the Shi'as, who celebrate this festival, especially; red is, as in Catholic cult, the emblem of the martyr.

168. Measure of distance, the value of which differs from province to province. However, it is of 42 for one degree. Hamilton, *East-Ind. Gazetteer*, vol. II, p. 722.

169. See the excellent work by M. l'abbé Dubois, entitled *Moeurs et institutions de l'Inde*, vol. I, p. 208.

170. *manjholi*, is a carriage with two wheels: this word is not to be found in any Hindustani dictionary: however, there are *maj-*

holi and *mjheli* (*majhili*), which are translated as a small cart. *Rath gari* is a carriage with four wheels.

171. *langui*, piece of cloth, with which Indians cover the middle of the body. We know that most of them possess only this dress.

172. The same practice is followed for the robe of Muhammad in Constantinople. See M. d'Ohsson, *Tableau de l'Empire ottoman*, vol. II, p. 391, ed. in-8.

173. For this water see the work entitled, *Les oiseaux et les fleurs*, p. 180, and the following article.

174. *andhi*, Hindustani word, which is synonymous with the Arabic *tufan*, which in India and particularly, in Bengal, has the sense of *ouragan*. The word *andhi*, which is understood by everybody who has heard Hindustani spoken or has read a few lines of this language, has very much embarrassed the late M. Langlès, and I do not know why, since he could have found very easily this word in any of the many Hindustani dictionaries that he possessed. This is how he expresses himself on the subject of this word in a note on *Voyages de Hodges*, vol. II, p. 142.

'*Aoundy, ouragan*. I do not know the origin of this word, on which none of my investigation has furnished any information. I am tempted to believe that there has been some error on the part of M. Hodges; because many scholarly travellers, whom I have consulted, admit that they do not know this word, and do not recall having heard this word in India: perhaps this is a corruption of the French word, *ondée*.'

175. *deo*, evil genie, the *pari*, or fairies are good genies. The preceding piece is an extract from *Bara Masa*, p. 29.

176. There is a tribe in Kabul, which calls itself *Khwaja Khizri* (*Ain-i Akbari*, vol. II, p. 164).

177. *Exodus* vi, 25: nos., xxv, 13 etc.

178. In the *fatiha* in his name he is addressed as *khwaja khizr mehtar ilyas*, Khwaja Khizr, the illustrious Elijah.

179. *rawayat khizr sun ponhchi hai mujhko.*
ke uska khat hai moje abe yaqut

180. *Bara Masa*, p. 62.

181. This month, which commences on 9-13 August and comes to an end on the same dates in September, is the last month of the rainy season. In the same season of the year the Egyptians perform on the Nile ceremonies, which are analogous to what has been described here.

182. In the text we have *shabe juma'h*, which is probably a synonym of *juma'h rat*, expression used in India for Thursday.

183. *khus abe hayati sati yu lab hai labalab*
    bire baghair is lab kon utar kon sakaiga.

184. This word is used in the sense of boat by the Gypsies or Bohemians, whose language appears to have been derived from Hindustani. See the investigative and interesting Memoir of Colonel Harriot on the oriental origin of the gypsies (*Transactions, R.A.S.*, vol. II, p. 518 sqq).

185. Shakespear, *Dictionary*, pp. 168 and 387.

186. *Transactions, R.A.S.*, vol. II, p. 539.

187. Old capital of Bengal, situated on the Ganges.

188. *Voyage pittoresque de Hodges*, Langlès trans., vol. I, p. 80.

189. Hamilton, *East-India Gazetteer*, vol. II, p. 192.

190. *Hidayat ul-Islam*, p. 270.

191. *Bara Masa*, p. 64. From among those which I have been able to consult, this is the only work which mentions Goga.

PART TWO

Saints Without any Special Fêtes

Abdulqadir

This saintly person, nicknamed Ghaus ul Azam, the great contemplative,[1] was born, according to Afsos,[2] at Jil, near Baghdad in 471 (1078–79) and received the robe of religious initiation from the hands of Shaikh Abu Sayeed. He was endowed with great virtues and the gift of miracles. Full of faith in him, a great number of people became his disciples and thousands of them were, through his mediation, instructed in the esoteric doctrines of the religion.[3] Even now, a large number of people recognize his saintliness and have great devotion for him. He is called Shaikh on account of his knowledge and virtues, although he was a Syed that is to say, one from the race of Husain. He lived for more than 90 (solar) years and set out for the eternal home in 571 (1175–6.)

Abdulqadir has written a number of well-known books on mysticism.[4] I believe he is the same person as mentioned in the article on Mo'inuddin above.

Sarwar

Sultan Sarwar, son of Syed Zainul Abdin,[5] gave himself up at a very tender age to piety and abstinence; and he was hardly an adolescent when he attained great purity of heart. Being obliged to fight a troop of idolators in the town of Balutch,[6] he became a martyr along with his brother. His wife died of sorrow and a young son followed them and thus all of them were buried in one grave, which is called the tomb of the martyr.

It is said that once a merchant was coming from Qandhar in Multan and his camel broke its legs near the saint's tomb. He did not know how to transport the animal's burden and he prayed to God at the tomb, and at once the camel's leg was reset. The merchant made an offering to the saint and having loaded the animal went his way and since then Sarwar's tomb has become a place of pilgrimage. It is said that three persons, a blind man, a leper and an impotent man went there and had the good fortune of being cured of their infirmity, by God's grace. These miraculous cures increased the faith in Sarwar and in the beginning of the winter season, people from all directions and from distant places come to make all kinds of offerings at his tomb.[7]

Twelve *kos* from Sialkot, in the province of Lahore, there is a place called Dhonakal, which is consecrated to Sultan Sarwar. All the year round, Muslims go there for pilgrimage but specially in the two months of the summer season, men and women come in large crowds from all over the province to place offerings.[8]

Daryai

Shaikh Shamsuddin Daryai, celebrated for his marvels, is buried in Depal-Dal, in the province of Lahore. Among the many miracles attributed to him there is one about a person called Depali, who was an orthodox Hindu but at the same time Daryai's disciple. Once he sought Daryai's permission to go on a particular occasion, to bathe in the Ganges along with his co-religionists. The saint advised him to remind him of his desire on the very day fixed for the sacred bath. Depali did this.'Close your eyes,' Daryai asked him. Depali closed his eyes and he found himself on the bank of the Ganges and there he bathed in the company of his parents and friends. On opening his eyes he found himself in the presence of his spiritual guide and this surprised him greatly. When his co-religionists returned and found him there they thought he had come back earlier, but when they came to know how all that had happened, they were plunged in an ocean of admiration.

Another fact, still more extraordinary, is the following: a few years after Daryai's death, carpenters felled a seris[9] tree which grew near the saint's tomb and they cut it into small pieces for use in construction. Suddenly a terrible voice was heard, the earth began to tremble and the trunk of the tree stood up of its own will. Terror-stricken, the workers took to their heels and it was not long before the tree grew green again.

These miraculous events contributed not the least to the spread of devotion for him. Even today his tomb is a popular place of pilgrimage. Rich and poor, men and women, go there every Thursday, especially on that of the New Moon and place there offerings, persuaded by the belief that through this act they will get their wishes accomplished. What is most striking is the fact that the guardians of Daryai's tomb are the Hindu decendants of Depali. In vain did the Muslims wish to take over from them these functions; but they did not succeed and the state of affairs continued till the time of Alamgir.[10] I do not know what is the position now.[11]

Qutbuddin

He is one of the most famous and most venerated Muslim saints of India. The town Qutub,[12] where he lived, is named after him, so is the monument, Qutub Minar, minaret of Qutub, raised near the town.[13] This superb and majestic edifice, sung by several Indian poets, is wearing away very fast. Near Qutub's tomb[14] there are several beautiful houses forming a square with a pond in the middle. These houses belong to the present Sultan of Delhi and to the princes of his family, who come to visit the saint's tomb. The late Shah Alam and several others from the family of Timur are buried in the town of Qutub and the nominal reigning[15] Emperor Akbar II had ordered that he and his wife be buried here.

Khwaja Qutbuddin Bakhtiar Kaki, son of Kamaluddin Musa, was born in Farghana.[16] God chose to attract him towards Himself, even when he was of a very tender age. The Prophet Khizr[17] appeared to him and made the celestial light enter his soul. At the age of twelve, he saw in a dream Khwaja Mo'inuddin,[18] whom he accepted as his spiritual guide, and wishing to have the blessing of his company, he set out to meet him. At Baghdad, he met a number of saintly persons in whose company he gained many spiritual benefits. Then he came to Multan, and cultivated friendship with Baha'uddin Zakaria,[19] and having come to know that Mo'inuddin lived in the Empire of Shamshuddin Altamash,[20] he proceeded to Delhi. Moved by divine inspiration, Mo'inuddin, on his part, set out for this town. It was there that these two elects of God, who were already connected by a spiritual line, got to know each other at the physical plane and were able to communicate their thoughts. However, they did not stay together in the city for long. Mo'inuddin went back to Ajmer, while Qutbuddin stayed on in Delhi, where through his mediation, a large number of people could participate in the profusion of divine grace. It was here that on the 14th of Rabi'ul auwwal, 630 (29th December, 1232) he left this perishable world and set out for the abode of eternity. His tomb is three *kos* from the city.[21]

Qutbuddin's tomb is frequented by a large number of pilgrims; but as in the case of other saints of India, people go there more out of curiosity than from devotion. The following description given by the Indian poet Fa'iz of what he saw gives us not too flattering a picture of the kind of people who visit these places.

One day as I was passing by the tomb of Qutbuddin I saw a vivacious stall-keeper, gentle like a bayadère and beautiful like a houri. She sold bhang,[22] beer and wine; her eyes brought trouble into hearts... There was a surprising crowd of people there... guitars and violins resonated from all sides; everywhere intoxicating drink was being sold. The crippled stood erect like candles. Many low-born people and slaves, whose ears were carrying the buckle of servitude, conversed peaceably among themselves, while others under the influence of liquor would fight with fists and feet and would before long draw daggers. The beautiful stall-keeper, who had attracted my attention, wanted to run way from this disorderly scene, but was cruelly assassinated and the full moon of her beauty, which was at its zenith, vanished in the perigee of death... Everybody was deeply moved by this sad event and some people were duped by their curiosity but several infamous blackguards were killed.

'O Fa'iz! Run away from despicable people, live day and night in the company of the good.'

Zakaria

Shaikh Baha'uddin Zakaria, son of Shaikh Qutbuddin Muhammad and grandson of Kamaluddin Quraishi, was born at Kot-Karor[23] in 565 (1169–70). Even as his father left this earth when he was a child, he continued to occupy himself with the spiritual science and before long attained the stage of excellence. Then with a wish to travel, he crossed over to Iran and Turan and reached Baghdad, where he attached himself to Shahabuddin Suhrawardi.[24] After having been his disciple for several years, he became his spiritual heir, so much so that Shaikh Iraqi and Mir Husain drew religious benefit from him. Then from Baghdad he came to Multan to live there. There too many distinguished persons received spiritual benefits from him. It is said that there was a very close friendship between him and Fariduddin Shakarganj.[25] For a long time they lived together in the same place. At last, on Safar 7, 665 (7 September 1266), a *pir* from Turan brought a sealed letter addressed to him and gave it to his son, Shaikh Sadruddin, who hastened to deliver it to his father. As he was reading it, Zakaria returned his soul to his Creator. A cry in unison rose from the house: 'the friend has joined the friend.'

Several miracles are attributed to this saint and it will take too long to recount them here. He is buried in Multan and his tomb is a place of pilgrimage.

His son, Shaikh Sadruddin, succeeded him in spiritual dignity and like his father he trained a large number of disciples among whom were many distinguished for their saintliness and virtue. He himself left this perishable earth in 709 (1309). Shaikh Ruknuddin,[26] his son, followed in his father's footsteps and after his death was buried like his grandfather at Multan.[27]

Fariduddin

Fariduddin Shakarganj, son of Shaikh Jalaluddin Sulaiman and descendant of Farukh Shah Kabuli, was born in Ghanawal, near Multan. When barely an adolescent, he went to study at Multan. There he benefited much in the company of Khwaja Qutbuddin Bakhtiar Kaki. He came to Delhi with his spiritual guide and, full of ardour, he entered the spiritual life. Some people say that at the command of the aforesaid saint he first went from Multan to Qandhar and Sestan,[28] and after having gained the necessary knowledge there, he came to Delhi where he became a disciple of Qutbuddin. It was there that he renounced the desires of the senses and gave himself up to cruel mortification and the painful practices of devotion. Then, after parting company with the director of the way of salvation, he returned to Hansi[29] where he led a quiet life till the former's death: during this time he went again to Delhi in order to collect the mantle and staff[30] of the order which Qutbuddin had taken over from his own master and which at the time of his death he had desired should be given to Farid. Carrying this precious trust, he left this town and went to reside at Patan,[31] where a whole world obtained heavenly favour through his mediation. He died in this town on Saturday, Moharram 7, 667 (15 September 1268) and was buried there.

Everybody knows that where Farid's eyes fall, the piece of earth is turned into sugar. This is the source of the nickname, Shakargunj, treasure of sugar, given to him.[32]

Qalandar

Shaikh Sharafbu Ali Qalandar was born at Panipat,[33] 30 *kos* north-west of Delhi. At the age of forty, he came to this city and had the benefit of being introduced to Khwaja Qutbuddin.[34] However, he devoted himself to exoteric sciences only for twenty years. Finally, the divine light shone on the mirror of his heart and he threw all his books into the Jamna and set out in search of religious instruction. In Asia Minor he derived great benefit from the company of Shams Tabriz[35] and Moulvi Roum,[36] and several other saintly persons. Then he returned to his country and lived permanently in the corner of his retreat till God decided to call him back. A large number of people have been eye - witnesses to his miracles, and even now his tomb is visited by a large number of pilgrims.[37]

If one believes M. W. Hamilton,[38] this famous Indian Muslim saint died in 727 (1323–4), but if at the age of forty he was in contact with Qutbuddin, who, as we have seen, died in 630 (1232–3), the date given by Hamilton may not be correct, because it would mean that Qalandar was more than 130 at the time of his death.

As a young person Akbar II, the so-called Emperor of Delhi, was taken to the tomb of Qalandar by his unfortunate father, Shah Alam, who consecrated to the saint a lock of his hair. This requires that for a certain period of time, the portion of hair one has cut is retained and is not touched. One should come to the same place again to get the hair cut and to place it there. It is said that the King wanted to complete the ceremony very much, but since this required a lot of money, and this could have created a lot of problems for him, he has been persuaded to defer this ceremony till now.[39]

The following are the words of the *fatiha* for the saint, according to the Muslim prayer book,[40] printed in Calcutta.

For the sake of the Prince of contemplatives, the chief of spiritualists, the illustrious Shah Sharfbu Ali Qalandar (May God sanctify his precious tomb) and also through the pure soul of Shah Sharfuddin Yahya Maneri, of Ahmad Khan and Mubarak Khan (May God sanctify their tombs), may God the Very High be pleased to accept the gifts and prayers offered to him.

Having thus pronounced his intention, the faithful will recite the first chapter of the Quran, then the verse of the Throne,[41] three times, the 94th chapter, three times; the first, three times; the 112th, ten times; the prayer *durood*,[42] ten times.

Awliya

The Prince of Shaikhs, Nizamuddin Awliya, son of Ahmed, son of Daniel, was born in Ghazna in 630 (1232–3). When he attained the age of reason he went to Badaun,[43] and there devoted himself with great success to external sciences. As he almost always won in debates with his fellow-disciples, he was nicknamed Conqueror of the Assembly, *mahfil shikan*. At the age of twenty he went to Ajodhan,[44] where he had the good fortune of becoming the disciple of Fariduddin Shakargunj,[45] who taught him the inner sciences. After leaving that place he went to Delhi to assume the spiritual leadership of the people. In his presence a large number of people given to the search of religious truths found great comfort. Among them one may cite Shaikh Wajhuddin of Chanderi,[46] Shaikh Chirag-i Dehli,[47] Shaikh Alaul Haque, and Raji Siraj of Bengal, Yaqub and Kamal of Malwa, Hasamuddin of Gujerat, Shaikh Burhanuddin and Khwaja Hasan from Deccan, Amir Khusro of Delhi, the respected Moughis of Ujjain, Ghayas of Dahar,[48] etc. His descendants and spiritual heirs continued to guide their co-religionists on the path of God till the time of Aurangzeb but after that period we do not know about his spiritual line.

The date of Nizamuddin's birth given by Farishta differs from what I have given above. According to him, the contemplative's father came from Gazna to Hindustan and lived in the town of Badaun, where our saint was born in Safar 634 of Hijri (October 1236). He was hardly five years old when his father, an extremely esteemed person, took the path to eternity. His mother took great care of him and brought him to Delhi when he attained the age of discretion. It was in this town that he learnt what children are ordinarily taught.

Nizamuddin entered paradise on Wednesday 18th Rabi'ul auwwal, 725 (4 March 1325) and was buried a little distance from Delhi, where one can still find his tomb near that of Khwaja Qutbuddin.[49] On account of his great piety, this friend of God was considered to be one of the most eminent saints of Hindustan. The chain of his religious initiation goes back to Shaikh Abdulqadir Jilani.[50]

Kabir

Kabir was one of the famous Hindu unitarians, who was venerated by Muslims as much as by his co-religionists. He established a new sect, Kabir Panthi, or partisans of Kabir, from whom Nanak, the founder of the sect of Sikhs, borrowed his religious ideas and propagated them with greater success.[51] Afsos says:

> If one believes in what so many people say, it is in Ratanpur in the Kingdom of Oudh, where the tomb of Kabir, the weaver, is to be found. This widely-known person, who lived during the reign of Sikandra Lodi,[52] spent a long time of his life in Banaras, engaged in the practice of piety. *Faqirs* consider him an orthodox and perfect person. He always recited verses[53] which he himself composed and which exude knowledge and love for God.[54]

During his lifetime, as after his death, he was venerated equally by Hindus and Muslims. Brahmans wanted to burn his body and Muslims to bury it, but

the legend says that his body disappeared at this junc-
ture.[55]

Lal

Baba Lal, also a Hindu, was a dervish who lived
in Dhianpur, in the province of Lahore. He could
argue with eloquence and facility and he employed
this talent to develop the immutable principles of
the Unity of God and in explaining other divine
attributes. People would rush to him and feel an
incredible pleasure in listening to him. He has left
behind him a large number of Hindustani verses
on religious themes, which many people recite
daily as a duty. This saint is widely respected
among nobles as well as by the ordinary people. It
is said that Dara Shikoh, eldest son of Shah Jahan
and brother of Aurangzeb, came to see Baba Lal
often and discussed things about God. In fact,
Munshi Chandra Bhan Shahjahani has written in
Persian a book which contains the pious conversation between these two persons.[56]

As in the case of Kabir, Baba Lal is thought to have
founded a Hindu sect, called Baba Lali.[57]

Daula

Shah Daula, the soul of contemplatives, was
originally a slave of Kamaiandar Sialkoti,[58] but the
friendship of *faqirs* made his life happy. He often
went to see Syed Nadir and enjoyed his edifying
company. Guided by celestial favour, Nadir threw
his last glance at him while he was dying. At once,
Daula entered a new state, his inner sight was

purified and he was able to see the spiritual light. He left Sialkot and took up his abode in Chhoti Gujerat,[59] where he built reservoirs,[60] wells, mosques, bridges and embellished the town which till then was not very flourishing. He got a very solid bridge built 5 *kos* from Amnabad on the river Dek, on the highway to Lahore, and this act was of great benefit to innumerable people. His generosity was such that had he been Hatim's[61] contemporary, people would not have mentioned the latter's name. People from far and wide would come to offer him presents, gold, foodstuff, etc., but he would return them gifts two or four times more in value. In the seventeenth year of the reign of Alamgir,[62] this saint returned his soul to God and was buried near the town which by his residence he had made prosperous and which a large number of people visit even today.[63]

Zahoor

Syed Shah Zahoor was a man of great under-standing and piety, and no *faqir* could compare with him so far as renunciation and austerity are concerned. He built near Allahabad a very small clay monastery which exists even today.

He liked to execute the most difficult exercises of devotion, such as reciting prayers in reverse.[64] His saintliness raised him above all his contemporaries and his miracles made him famous. I have heard the following from my father. The late Nawab Um-datulmulk Amir Khan, Governor of Allahabad, was suffering from a dreadful chronic disease. He con-sulted the most skilful doctors, but they could not cure him. One day a nobleman praised Shah

Zahoor to him and so the Nawab felt the desire to see the contemplative and prayed to him to come to his place. As he entered the Nawab's house he pronounced these words: 'The prayers of *faqirs* attract God's mercy; their presence drives away unhappiness.' At once the malady declined in intensity and the Nawab felt relief from pain. After a few days the Great Hakim responded to the great saint's prayer and gave to the Nawab a perfect cure. It is not necessary to trust only the remedies, the prayers of *faqirs* are at times more efficacious.

Shah Zahoor was an Imami and belonged to the spiritual line of Chishti.[65] His excellent masters had been contemplatives especially so Syed Shah Fath Muhammad, who was a distinguished person both in the exterior and interior sciences, and a famous person of that century. A large number of people recognize his saintliness and speak about his supernatural powers. I have heard many things about this saint from Mian Shah Ghulam Rasul, a direct descendant of Shah Zahoor Muhammad and a very religious and truthful person. (However, I do not know if Ghulam Rasul is still alive or about who holds the highest spiritual position, *sajjada nashin*, in his line). I was born when Shah Fath Muhammad was alive. It is said that he claimed to have attained the age of 300 years and had seen the construction of the fortress of Allahabad, a claim which many believe to be true. It is possible that God wished that in this last phase of Time, a person of such an extraordinary nature be born in the family of the Prophet[66] and that he should live for such a long time. However, what is certain is that this distinguished person was alive till recently. My father, who had the honour of seeing him several times, testified to the saint's miracles and often talked of the efficacy of his amulets. This servant of God was really full of moral qualities and had put on the robe of spiritual poverty. But as we always finish with death, and the gain of life is nothing but death, he terminated his existence in Allahabad. We do not know about any sect founded after him or his descendants, spiritual or temporal.[67]

Hazin

In Banaras there is a large number of Muslim tombs among which is the distinguished one of Shaikh Muhammad Ali Hazin Gilani.[68] This saintly person had during his life time, got his grave prepared and often on Thursdays[69] he would go there and sit near it and distribute alms. 'He sees without horror the approach of death who takes it as an entry to Immortality: death cannot change the state of a man, who has known death in his life time.'[70]

The Shaikh combined the knowledge of exterior with the interior sciences. His skill in writing prose and verse was the least of his merits.[71] He was the best among the authors of his time and he should be a model for writers of our times. He came to Hindustan during the reign of Muhammad Shah. After having stayed in Delhi[72] for a few years, he went to Banaras, where he lived in extreme solitude, from where he never went out to see people, whether great or small and far from receiving anything from anybody, he gave to the poor according to his means. His life was always irreproachable; he had no desire but to be united with God. Things were revealed to him and he had been endowed with the gift of working miracles; it is said that the Sun was under his command, and if he wanted he could execute other marvels not less extraordinary.

Everybody knows that this contemplative without hypocrisy, was far from being a person to have advised the Nawab of Awadh, Sujaudaullah, to attack the English: on the contrary, he wisely advised him to live in peace with them. He died in 1180 (1766–67), after the rout of Buxar[73] and went to live in paradise.[74]

Nothing would have been easier than extending this *Memoir* by talking of several other venerated saints of Muslim India who have become celebrities. In the Hindustani books that I have consulted there are references to more than one hundred *pirs* about whom one should know. But having neither the ability nor the wish to talk about all those who merit mention, I have restricted myself to a few. I have written especially about twenty persons and have added notes on a number of them more or less their equals. I believe this is enough and I shall apply to myself what Mir Taqi[75] has said:

> You might have many things yet to say, however important they might appear; but the seal of silence is at present preferable and it is better to renounce speech.
>
> *Kitni wus'at teri [sic] bayan men hai*
> *Kitni taqat teri zuban men hai*
> *Lab par ab muhre khamushi bihtar*
> *Yan sukhan ki faramushi bihtar.*

NOTES

1. 'By this is understood the unique person, on whom God turns His regard, eternally. He is the pole (he is also called *kutb*, pole), who fills with the spirit of life both the lower and higher nature'—See the notice of M. de Sacy on the work entitled *Tarifat* (*Not. et Extr. des Mss.*, vol. X, p. 81).

2. *Araish-i Mahfil*, p. 61.

3. That is to say, Sufism, *tasawwuf*.

4. *Araish-i Mahfil*, p. 62. On one of the mystical treatises of this celebrated person, there is a commentary written by Abdulla Husain Kes-diraz of Kalbargah [*sic*] in the Hindustani dialect of the Deccan. This work is cited in the catalogue of the library of Tipu by M. Ch. Stewart and in the Catalogue of manuscripts of the library of Fort William College of Calcutta; it is entitled *Nashatul ishq*, that is to say, the pleasures of love (divine).

5. The tomb of this saintly person is at 4 *kos* from Multan: people from all directions come here for pilgrimage during the summer season and stay here for a few days. I do not know if this Zain-ul Abdin is the same person as is mentioned in *Ain-i-Akbari*, vol. II. p. 152.

6. Apparently, Kelat, their capital. See Hamilton. *East-India Gazetteer*, II, p. 81.

7. *Araish-i Mahfil*, p. 165.

8. Ibid., p. 184.

9. *Mimosa seris.*

10. Probably Alam Gir II, who reigned from 1733 to 1756.

11. *Araish-i Mahfil*, p. 75.

12. See for an exact description, Hamilton, *East-India Gazetteer*, vol. I, p. 473. M. C. Elliot has recently given to the Asiatic Society of London, a copy of the inscription written on this tower. If we are to believe Bernier (*Voyages*, vol. II, p. 75, Amsterdam, 1723), this edifice was once a *dehra*, or temple of idols, and the inscriptions are in an unknown script, different from that of any Indian language.

13. *Qutb* is not taken here in the mystical sense it sometimes assumes and which M. de Sacy has explained satisfactorily in his translation of *Pendamah* or *Attar's Book of Counsel*, p. LVIII. It is used for Qutbuddin, which is the honorific title of the saint and which means the pole of the religion. This is quite close to the Turkish practice, where one would say, Baqi, instead of Abdul Baqi, Servant of the Eternal.

14. Hamilton, *East-India Gazetteer*, vol. I, p. 473.

15. In the eyes of the natives of India, the Englishmen administer the country under the order of the Great Mughal; they are considered his lieutenants or vizier. Afsos says in unambiguous terms, Hindustan has for sometime been dominated by a multitude of small sovereigns, who fight over each other's possession. None of them recognizes the legitimate authority of the Mughal, which they should have done, except the English *Sahabs* (messieurs), who have never ceased to be obedient, so much so that in 1222 (1897), they recognized the supreme authority of Akbar Shah, son of Shah Alam (*Araish-i-Mahfil*, p. 211).

16. Country and town in Transoxonia.

17. See in the first part, the article on this prophet.

18. See the article on this saint.

19. See the following article.

20. Pathan Emperor of Delhi, who reigned from 1210 to 1225.

21. That is to say, the city of Qutb or Qutub, as we have seen above.

22. We have already talked of this intoxicating liquor.

23. City in Multan.

24. Celebrated contemplative, author of several famous mystical works; he was born in 539 (1144), and died in 632 (1234). See the Notice on the lives of the Sufis, including that of Jami, by M. le baron Silvestre de Sacy, in vol. XII of the Notices on Manuscripts.

25. See the article on him.

26. This Shaikh had the patronymic surname, Suhrawardi, as we have seen above in the article on the month of Rajab. His grandfather Zakaria had been the disciple of Shahabuddin

Suhrawardi and apparently he had taken his surname and transmitted it to his descendants. See *Preliminary Observations.*

27. *Araish-i Mahfil*, p. 164. Also see *Ain-i Akbari*, vol. II, p. 113, and Hamilton, *East-India Gazetteer*, vol. II, p. 242.

28. A large province of Baluchistan.

29. A city in the province of Delhi.

30. See *Preliminary Observations.*

31. This name, which is common to cities of India, refers, here, to a city in the province of Multan, also called, Ajodan, situated in the *sarkar* or district of Debalpur. *Ain-i Akbari*, vol. II, page, 286.

32. *Araish-i Mahfil*, p. 166.

33. It is near this city that between the Muslims and the Mahrattas the Battle of Panipat was fought in 1761 and which was won by the former and has been celebrated in a Hindustani poem entitled, *Jang-nama*, that is, *Book of War*. Mackenzie, *Collection*, II, p. 145.

34. See above, the article on this saint.

35. That is, Shamshuddin Tabraizi, celebrated Persian poet. M. Jules Boilly, the distinguished painter has, in his beautiful collection of Persian manuscripts, a correct copy of the *diwan* of this poet, the original of which belonged to Scheidius.

36. Maulawi Jalaluddin Rumi, great Muslim spiritualist, founder of the order of Maulwi and author of a very famous poem, known under the imprecise title, *Masnavi*. During the time mentioned, he lived in Cogni (Iconium). (D'Herbelot, *Bibliothèque Orientale*).

37. *Araish-i Mahfil*, p. 64.

38. Hamilton, *East-India Gazetteer*, vol. II, p. 367.

39. Ibid.

40. *Hidayat-ul-Islam*, p. 269.

41. That is, versets 255–8 of the second chapter of the Quran.

42. See *Doctrine et devoirs de la religion musulmane*, p. 222.

43. A city in the province of Delhi, which is remarkable only for its antiquity. *Ain-i Akbari*, vol. II, p. 87. Hamilton, *East-India Gazetteer*, vol. I, p. 291.

44. A city in Multan, which has already been referred to.

45. See the article on this saintly person.

46. A city in Malwa.

47. Nasiruddin Chiragh Dehli, that is, the Help of Religion, Lamp of Delhi, buried in this city (*Araish-i Mahfil*, p. 166; *Ain-i Akbari*, vol. II, p. 87). There is another saint with the same name Shah Nasiruddin, who is buried in Jalindar, city in the province of Lahore. A very large number of pilgrims come here, especially during summer and offer at his tomb gifts and their wishes. *Araish-i Mahfil*, p. 172.

48. The old city of Malwa, which has been the capital of this province.

49. See *Ain-i Akbari*, vol. II, p. 87.

50. See the article on this saintly person. The preceding notice is an extract from *Araish-i Mahfil*, p. 60.

51. H. H. Wilson, 'A sketch of the religious sects of the Hindus' (*Asiatic Researches*, vol. XVI, p. 53).

52. King of Delhi, of the dynasty of Pathan or Afghan Lodis, who ruled from 1488 to 1516.

53. *Dohra*, a Hindustani word which is synonymous with the Arabic *bait*. The savant, M. Wilson, has included translation of several verses of Kabir in the excellent Memoir on the sects of the Hindus, which has enriched vol. XVI of *Researches asiatiques*.

54. *Araish-i Mahfil*, p. 92.

55. *Ain-i Akbari*, vol. II, p. 16.

56. *Araish-i Mahfil*, p. 176.

57. *Asiatic Researches*, vol. XVI, pp. 26 and 53.

58. The original Persian work which is the base of the book by Afsos carries these words, *Kahaiderah sakin Sialkot*.

59. *Chhoti Gujerat* or small Guzaerate.

60. To be correct, ponds, *talab*.

61. This Arab, who is famous for his generosity, is the hero of a Persian novel, which has been recently translated in English by M. Forbes, the painstaking and esteemed Orientalist. There is also a Hindustani translation of this work, which has the

emphatic title *Araish-i Mahfil*, that is, the Ornament of the Assembly, the same title as that of the book by Afsos, which I have often cited—*Catalogue manuscrit des livres hindoustani, persans et arabes du Collège de Fort-William à Calcutta.*

62. Better known by his other title of Aurangzeb , that is, Ornament of the Throne. The seventeenth year of his reign corresponds to the year AD 1675.

63. *Araish-i Mahfil*, p. 185.

64. Extraordinary practice of piety. See Golius, *Lexicon arab. lat.*, p. 2453, for the word *naks.*

65. *Silsila-e Chishtia.* See, in the first part, the article on Mo'inuddin Chishti.

66. Syeds are from the family of Muhammad, and the descendants of Husain.

67. *Araish-i Mahfil*, p. 83.

68. In other words, from Gilan, since he was born in Hispahan in 1692; but because he belonged originally to that place and lived there for a long time. Belfour, *The Life of Ali Hazin, written by Himself*, pp. 50, 135, 169.

69. A day especially dedicated, as we have already seen, to the commemoration of the dead and to religious exercises for the peace of the soul of the dead.

70. *Jo baqa apni fina samjhai wai dukh bhartai nahin*

  Mar mitai jo zindigi main woh kabhi martai nahin

71. He has left behind collections of poetry or *diwans* and very interesting memoirs, which have been recently published in English by M. F. C. Belfour, with the title *The Life of M. A. Hazin, written by Himself*, and funded by Oriental Translation Fund.

72. It was there that he wrote his *Memoirs*, the details of which do not go beyond the period of his stay there. The book exudes his most fervent piety and gives a very favourable idea about Hazin. One finds in this book very liberal views about religion, which is at the same time in accord with the spirit of the Quran and the system of the Sufis. One finds there that he knew Christianity through the Bible and the Christian missionaries; but far from being converted he was strengthened in his faith.

73. Buxar, city in the province of Bihar, famous for the great victory the English had near it in 1764 against the united forces of Shujauddaulah and Qasim Khan, the Nawab of Bengal. Hamilton, *East-India Gazetteer*, vol I, p. 304.

74. *Araish-i Mahfil*, p. 88.

75. *Kulliat Meer Taki*, Calcutta, 1811, p. 910.

2

Customs of the Musulmauns: An Essay on Mrs Meer Hassan Ali's *Observations on the Musulmauns of India*

As portions of my *Memoir* on the special features of the Muslim religion in India were coming out in this journal, the book with the above title was published in London. With the thought that my own ideas, taken from the Hindustani writings which I have been able to know, should find confirmation in them, I read the book avidly, and I should say, with keen pleasure, because I found there reproduced the ideas which had been suggested to me by my readings and which I had later explained in my book and also certain explanations which I had looked for unsuccessfully. In the framework of this book which is larger than my *Memoir*, the questions I had treated have naturally found a place, some of them have been developed at greater length while others have been ignored; thus, really, most of the information I have given on the Muslim saints of India remains the most exhaustive so far in this genre.

The author of this book was, more than anybody else, in a position to make exact observations on Indian Muslims. As the wife of a distinguished Muslim, she lived for twelve years in the midst of the family of her husband, without however being shut up in his harem. She was thus able to see things with her own eyes and the instructive information given by her husband and the father-in-law kept her on guard against her own illusions.

Her husband, Mir Hassan Ali,[1] is a highly educated Muslim, who lived for several years in England. He was attached to the military school of the East India Company at Croydon and it was there that he translated into Hindustani the portion of *The Vicar of Wakefield* which was published in *Muntakhabat-i Hindi* compiled by the scholar Mr. Shakespear. He married in England the lady who has written the *Observations* and brought her to India where she instructed herself deeply in the beliefs and customs of the Muslims in this beautiful part of the world.

The writer's father-in-law, Mir Hajee Shah, is presented by Mrs Hassan Ali as a very learned and, above all, a very religious person. Even as she is a good Christian, a fact of which she has given evidence at several places, she does not hesitate to compare him (vol. II, p. 422) to the Israelite whom Jesus Christ found *without any guile, without any artifice* (John, i, 47) and (vol. I, p. 146) following the example of Saint John, *and other sheep I have, which are not of this fold* (x,16) *in my Father's house there are many mansions (XIV, 2)* she says that at his death, *his pure soul was prepared to meet his Creator, in whose service he had passed this* *life, in all humility, and on whose mercy alone his hopes for the future were centred* (vol. I, p. 28)[*]

These two persons are Mrs Hassan's authority and they inspire effective confidence in whatever she says. What should please the impartial reader particularly, is that the author of these *Observations* is far from sharing the prejudice which people generally have against Muslims and their religion. She might rather be criticised for the contrary; but this criticism is also praise for the lady and for Muslims whom she has seen from close quarters for twelve years and who have deserved to be the object of her enthusiasm.

It is in a series of letters that she has reviewed the mores, the customs, and the religious opinions of Indian Muslims. The course she has adopted is not methodical, but the material has been linked so well through ingenious transitions, that far from being an incoherent piling up, they form one whole, full of interest and charm. I will limit myself to presenting succinctly the varied pictures which pass before the reader's eyes and at the same time taking care to elaborate a little more on the articles which are related to the questions which I have treated in my *Mémoire sur des particularités de la religion musulmane dans l'Inde* and I would also urge the readers of *Journal asiatique* to consider these pages as a sort of appendix to my work.

In her first letter Mrs Hassan Ali deals first with the characteristic simplicity of the mores of the Muslims, their charity and then with the Syeds, the fast during Moharram and its origin, etc. In the midst of

---

[*] Instead of retranslating, I have retained, throughout, Mrs Ali's own words (Translator).

these interesting details with which the letter is full, one reads on p. 22 an account, which I believe, needs to be quoted:

> Amongst the number of Hosein's brave defenders was a nephew, the son of Hasan; this young man, named Cossum, was the affianced husband of Hosein's favourite daughter, Sakeena Koobrah; and previous to his going to the combat on that eventful day, Hosein read the marriage lines between the young couple, in the tent of the females.

These details were not known to me, when I drafted my *Memoir* and, therefore, I had given a bad rendering of a very vague passage from *Bara Masa*, where an allusion to this fact has been made. The passage in vol. VIII of the *Journal*, p. 165 and p. 34 of the special edition, should then read as follows:

> It is reported that at the moment of his death, Husain wanted, in conformity with the last wishes of his brother Hasan, to unite Kasim, the latter's son to his beloved daughter. He therefore clothed the boy in the nuptial dress which is appropriate for a son-in-law and pronounced the words which are used for the celebration of marriage.

Mrs Hassan Ali then says that Husain's head was cut off and taken to the barbarous Yezid, but that one of his wives prayed to him to return it to her and she took it to the family of Husain who were at that time the Caliph's captives. They took the head to Karbala, where it was buried in the grave which contained Husain's body, forty days after it was beheaded. This account explains the fake miracle described by the famous traveller Chardin, which consisted of the reunion of the head and body of Husain. The miracle, he says, is celebrated by the Persians as a festival called *sarotan*, head and body. But this celebration which I have referred to in a note in my *Memoir*[2] is simply a commemoration of the event which I have just mentioned.

In her second, third and the fourth letters Mrs Hassan Ali gives a detailed description of the manner in which the festival of Moharram is celebrated and makes the observation, with reason, that this festival is contrary to the spirit of the Quran. She describes the form of the *ta'zia* (the representation of Husain's tomb), she speaks of the material used in making them, which could be anything from silver to talc, bamboo and paper. The most beautiful one that she ever saw was made in England and belonged to the King of Oudh. Costly *ta'zias* are kept in Imambaras, others which have no value are buried along with the decorations in Muslim graveyards, which in India, according to Mrs Hassan Ali, are called Karbala, the name given to the field where Husain and his companions were killed and buried. The above should be a correction of the article in *Hindustani Dictionary* by Taylor, which was reproduced by Shakespear in his own dictionary and by Mr Smyth in the abridgement that he has given of the former. From the summary of Mr Smyth's article which I have given in my *Memoir*[3] it would appear that, in order to bury the *ta'zia*, people go sometimes to the *dargah* or tomb of Husain at Karbala, while as a matter of fact, Karbala in this context refers to the cemetry of any town, in which the festival is held.

Mrs Hassan Ali tells us that the elegiac hymn called *marsiya* (not *musseah* without an *r*) which I have mentioned in my *Memoir*[4] is a poetic composition of high literary merit. It is written in the Hindustani language and narrates events the solemnity of which is recalled during Moharram.

While describing the procession, which is taken out on this occasion, she mentions the banners which people carry and observes that they are blue, purple, green, yellow, etc., and that they are not red, since that is the colour of the Sunnis. I doubt the correctness of this observation, since the colour of the Sunnis is not red, but black,[5] while red is for Indian Muslims, as in the Catholic religion, the emblem of the martyr.[6]

In the rank of the procession, one can see a man in mourning carrying a black pole from which are suspended two naked swords attached to a reversed bow; this man represents Abbas Ali [*sic*], a relation of Husain and his standard bearer. In a magnificent building in Lucknow is preserved the knob of the banner which he carried on that sad day; and at the time of Moharram, as I have said, the banners are touched to this relic. The *fatiha* for this saintly person written in Arabic, Persian and Hindustani can be found in the Muslim prayer book printed in Calcutta under the title, *Hidayat ul Islam*, p. 274.

Husain's horse, called Duldul, is sometimes represented in the Moharram procession by a beautiful white horse caparisoned in the ancient Arab style. They place carefully on it a blood-soaked covering, in order to give an idea of the animal's suffering; its legs have patches of red and arrows are fixed on different parts of its body in such a way that it appears to be pierced. On its saddle is placed an Arab turban with a bow and arrows.

On the seventh day of Moharram takes place the commemoration of Qasim's marriage, which has been mentioned above, a ceremony which is described only in Mrs. Hassan Ali's book. It is called *menhdi*, a word which comes from the name of the plant (*lawsonia inermis*) called hinna in Arabic and whose powdered leaves are used by Indians to redden the feet and hands. One should know that it is a custom in India that a day before the marriage, the bride's father carries on stretchers and with great pomp to his future son-in-law's house, powder of *menhdi* put in silver vases which are decorated with coloured paper and talc. It is precisely this ceremony which they wish to repeat on this day. They carry in the streets plates of *menhdi* and all the items that one offers at a marriage, such as sweets, dried fruit, garlands of jasmin, artificial flowers of talc, and other small artefacts, which are removed later. Then come the replica of the tomb of Qasim and two palanquins; the first apparently represents Qasim's and the second that of Sakeena Kubra, Husain's daughter and Qasim's fiancée. This procession is followed by bands of musicians who are accompanied by men carrying torches. The procession goes to an Imambara, a kind of funeral chapel of which a description has been given in my *Memoir*.[7] When the horse that represents Duldul arrives, it takes a round of the *ta'zia* and the things mentioned above are put there: they remain there till the tenth day of the festival, when they are brought again to the procession which goes to the cemetery or Karbala, for burial of the replica of Husain's coffin. On the night of *menhdi*, red and green candles are placed before the *ta'zia* in the Imambara; the red in memory of the martyrdom of Husain and the green to recall the poisoning of Hasan.

In Imambaras and other monuments raised by Indian Muslims, one can notice niches where they keep silver models of the Ka'ba (*temple de la Mecque*), of Husain's tent, of the tomb at Karbala, etc. I will not summarize Mrs Hassan Ali's fifth letter which is almost entirely devoted to the toilet of Indian ladies, a letter very interesting from the ethnographic point of view.

In the following letters up to the eleventh, Mrs Hassan Ali examines the principles and duties of the Muslim religion and all that distinguishes the sect of Sunnis from that of the Shi'as or Imamis, who are the most numerous in India. Since this subject has been the theme of a large number of books, these letters which deal with Islam in general have nothing very remarkable about them. However, one reads with interest what has been said about Imam Mahdi, who according to the Muslims, would appear at the end of time along with Jesus Christ, when Mecca would be full of Christians, and when the world would be converted to the faith of Jesus Christ, as the Christians believe. Only one error has crept into Mrs Hassan Ali's narrative: it was Abu Bakr, the Prophet's father-in-law, who succeeded him and not Umar. Usman succeeded Umar and Ali, the former.

About the Quran, Mrs Hassan Ali says that it is read only in the original: however, persons who do not know Arabic have books which contain a passage-wise commentary in Persian. There are also interlinear translations of the Quran in Hindustani; one of them was published in Calcutta in 1828.

With regard to the Qibla, the point to which Muslims turn in order to pray (Mecca), she quotes[8] the following from a commentator of the Quran, which is worth reproducing:

The Sovereign's Kiblaah is His well-ornamented crown.
The Sensualist's Kiblaah, the gratification of his appetite.
The Lover's Kiblaah, the mistress of his heart.
The Miser's Kiblaah, his hoards of gold and silver.
The Ambitious Man's Kiblaah, this world's honours and possessions.
The mere Professor's Kiblaah, the arch of the Holy House. And
The Righteous Man's Kiblaah, the pure love of God, Which may all men learn and practise.

Mrs Hassan Ali has explained what Muslims understand by the word, Gospel. This word does not mean either the four evangelists or even the New Testament, but only the words of Jesus Christ—the Sermon on the Mount, for example—and all the precepts which came from his mouth. Thus restricted to the discourses of our Divine Legislator, Evangel is found in a Life of Holy Prophets, often cited by Mrs Hassan Ali. This book, which has an Arabic title, *Hayat ul qulub*, or *The Life of the Hearts*, is like our Life of Saints. The book, originally in Persian, has since been translated into Hindustani.

A large number of Indian Muslims recite on 15th Rajab, a supererogatory prayer which is called David's Mother's Prayer and comprises sixteen large pages. They fast on this day and begin the prayer after taking a bath. Mrs Hassan Ali has not included a translation of this prayer; however, she mentions the miraculous event which caused the change in the name from 'The Opening of Difficulties' (*eplanissement des difficultés*) which it had at first to the one that is given to it now. The narrative being a little long, I would refer the reader to Mrs Hassan Ali's book.[9]

Some devout Muslims fast for forty instead of thirty days, while others fast for three months, beginning in the month preceding Ramazan and coming to an end a month after. These months are Sha'ban, Ramazan, and Shauwwal and not Rajab, Sha'ban and Ramazan as I wrote in my *Memoir*.[10] Some others fast all the year.

At the festival called in India Baqr-id, the Festival of Oxen—that is, of sacrifices—people make preparations to take to the *idgah* (and not *eade-gaarh* with an *r*), a sort of sacred abattoir, the animal meant to be slaughtered. They go there in a procession. The chief *mullah* recites the words of the prayer meant for this occasion; this prayer is included in my translation of the Muslim prayer book, p. 167 sqq. The *mullah* then offers the knife to the most eminent person in the procession, who with his own hands slaughters, in God's name, the camel which he wishes to offer as a sacrifice. This ceremony is announced with a salvo of artillery, which is a signal for the day's rejoicing. At Lucknow, this procession is nothing but a cavalcade consisting mainly of the king and the army, which follows him on foot and on horses. The elephants which are brought there are very well-washed, their hide is oiled, their heads are painted in bright colours, their trappings are very rich, their *hauda* (seat) is made of gold and silver and their drapery is of gold-fringed velvet. The horses are equally well-harnessed; stars are painted on their breast and back, while the tails and manes are tinted with *menhdi*. The soldiers have on their turbans the figure of the fish, the coat-of-arms of the Royal House of Oudh. The King's carriage is drawn by four elephants[11] whose height equals their corpulence; it is open on all sides, but over it there is a canopy in crimson velvet embroidered with gold thread. Servants, moving fans (*tchauri* or *thaunri, chunri*) stand on all sides of the king and they hold near him the regal parasol (*aftabi*, and not *afthaada* as it is always printed in Mrs Hassan Ali's book.) An unoccupied royal palanquin, *nalki* [sic] is followed by another *palki* and another vacant carriage drawn by eight black horses, equally empty. After these, the *harakaras* (messengers), the masseurs, *chobdar* (and not *chobdah*, without an *r*), dignitaries on elephants and notable men of the state. Once the ceremony is over the cavalcade returns to the palace; the king then takes his seat on the throne and receives gifts, *nazar* (and not *nazza* without an *r*,) which are offered to him by those present.

The most acceptable view about the saintly person called Khizr is that under his name Muslims honour the prophet Elijah, since in the *fatiha* for the saint, a translation of which I have included in my *Memoir*,[12] he is called Elijah, Ilyas, and Mrs Hassan Ali calls the *bera* or boat which is floated on this festival, the boat of Elijah, *ilyas ki kashti*. Another point made by Mrs Hassan Ali which supports this view is that Elijah had apparently received from God a special power over the element of which Khizr is patron. He actually closes and opens the sky[13] and his mantle also helped his spiritual successor Elisha to make a passage across the waters of the Jordan.[14]

Religious Muslims of India perform on the New Moon an interesting ceremony which is worth mentioning. They bathe and change clothes and when the discharge of artillery announces the New Moon, they take the Quran, open it on the page in which Muhammad has praised God for this particular good thing (i.e., the New Moon) and then they put on it a small mirror and hold the book in such a manner as to be able to see in the mirror the image of the moon. They recite the special prayer for the occasion and then they get up. Members of the family embrace each other and the servants salaam their masters. Each greets the other with '*May the New Moon be fortunate!*'

The Festival of the Dead, called Shab-i barat, which takes place on Sha'ban 14, is all the more important for Shi'as since it is the birth anniversay of Mahdi, the last Imam. Many illiterate Muslims believe that on this day trees talk to each other.

In her twelfth letter, Mrs Hassan Ali gives interesting details on the interior of the *zenana* or gynaeceum, and describes the mores and customs of Indian Muslim women. On this subject, she says that women are, contrary to the general belief, completely free and very happy in the *zenana*. Mirza Abu Talib Khan goes still further; he holds the view in his *Travels*[15] that they are really *freer and greater mistreses of their actions than the Englishwomen*. Since she has lived for twelve years in the midst of Muslim women, her evidence has a certain weight when she contends that the wives of the same husband have among them perfect accord, they love each other as sisters and have almost the same affection for the children of their companions as for their own.

The thirteenth and fourteenth letters with which the first volume comes to a close, deal with all that concerns marriage. Mrs Hassan Ali's description of the ceremonies which take place for this solemn contract is in accord with that of the late Colonel Mackenzie published in volume III of the *Transactions of the Royal Society of Great Britain and Ireland*[16] and with the poetic picture drawn by Hasan in his Hindustani poem entitled *The Magic of Eloquence*.[17]

The fifteenth letter deals with the birth of children and their education. We find here among other interesting details one on kite-flying, which is very popular in India. Children fly kites from roof-tops but they are not content with watching them sail tranquilly about in the sky. They attempt to hitch on to those of their friends' and allow, as it were, the kites to fight. In order to achieve victory over the other kites, they coat the string with a paste called *manjha*, made of soot and powdered glass, and in this way they succeed in cutting the string of the friend's kite and make it fall—to the exclamations of idle persons and other children in the street, who fight among themselves as if the kite was a precious booty. We find a poetic description of this popular game of young Indians in the Hindustani poem by Jawan, entitled *Bara Masa* or the Twelve Months, p. 82 of the Calcutta edition. The poet says, among other things, that schoolchildren often tear off leaves from their books to make kites and that the game is not confined to children, young and even old men occupy themselves with it. Finally, during winter especially, people indulge in this pastime. Mrs Hassan Ali ends this chapter with some reflexions on the administration of justice in Oudh, and remarks that debtors are not put in prison there.

The sixteenth letter deals with the professions and commercial activity of Indian Muslims. The most interesting pages are those which enumerate the cry of hawkers in the streets of Lucknow. From the noise of these cries, I have selected the following:

Sipi wala, applier of suckers.
Djonk or *kirah lagane* (and not *lurggarny* with an *r*) *wali*, the female applier of leeches.
Tail ke [*sic*] *achar wala*, seller of fruit kept in oil.
Mithai wala, seller of bonbon.
Khilaune wala, seller of toys made of wood, of talc paper, bamboo, clay etc.
Pankha wala, seller of fans.
Chirya wala, seller of birds, such as parrots, bulbul (a kind of nightingale), maina (coracia indica), etc.
Atash (and not *artush* with an *r*) *bazi*, fireworks. Their kind is infinite, and have different names in Hindustani.
Chabeni, roasted corn.[18]
Dahi khatti [*sic*], sour yogurt.
Mala'i, cream curd (sort of cheese with cream).
Barf wala, seller of ice or sherbet.
Menhdi, we have already spoken of this Indian toilet item.
Surmah (and not *sulmah*), a sort of a collyrium made of antimony.

At the end of the list I should point out that Hindustani is the only language spoken in Oudh, as well as in the provinces of Bihar, Allahabad, Delhi and Agra and a very large part of the Deccan also, although elsewhere it is used concurrently with the local language. In her seventeenth letter, Mrs Hassan

Ali returns to women, their pastimes, their games. She talks of slavery in India and of several other subjects.

In the eighteenth letter she deals with the physical inconveniences Europeans who live in India have to face. In her nineteenth and twentieth letters, she describes a journey to Qannauj, the old capital, and to Delhi, the modern capital of Hindustan. Her narrative is of the greatest interest. At Qannauj, she visited the *dargah* or the tomb of the Muslim saint called Makhdum, who has been mentioned by Hamilton also (*East-India Gazetteer*, vol. II, p. 74.) In Delhi, she went to offer homage to the nominal Emperor Akbar II and his Queen. Their Majesties extended to her a special reception, since she was the wife of a Syed.[19] Akbar's looks, says Mrs Hassan Ali, were venerable, his features very attractive; he appeared to be very intelligent; his conversation was amiable and relaxed. In manners he was equal to the most distinguished Europeans. His unfortunate position led him to make serious reflexions on the nothingness of the material world; and far from employing his time to intrigue in order to regain the empire of his forefathers, he spent it in exercise of the most fervent piety as a true dervish would and distributed all his income among unhappy people, instead of making arrangements for payment of some agents (i.e., political people.)

In Delhi, Mrs Hassan Ali also visited the tomb of Nizamuddin Awliya, the saintly person to whom I have consecrated an article in my *Memoir*.[20] This monument is architecturally like other Muslim tombs; it is square, with a dome of beautiful white marble, constructed by the pious monarch Akbar II about whom we have just talked. The cobbling as well as the columns are of marble. The tomb is in the centre of the edifice and is seven feet long and two feet wide and one foot above the ground. On the sides are engraved in black, versets from the Quran; it is covered with a golden fabric resembling a pall. This tranquil place is considered sacred by the Muslims. There one does not hear even the sound of steps because pilgrims are obliged to take off their shoes. The attendants of this tomb are religious people, who live on the bounty of pilgrims. They pass their time in the exercise of piety and read, among other things, the Quran over the ashes of the saint and they look after the property of the monument.

The twenty-first and twenty-second letters, which I shall skip, deal with the natural history of India. The twenty-third, twenty-fourth and twenty-fifth letters are about sufis, dervishes and other classes of religious Muslims, such as the followers of Syed Ahmad Kabir[21] and the disciples of Madar, who are called *dafali*, since they use in their religious ceremonies, small drums called *dafla*. To the details I have given on the pilgrimage which takes place at the tomb of the saint, I might add the following from Mrs Hassan Ali.

A descendant, real or fake, of Madar, waits near the tomb to receive *nazar* or offerings, the income from which is very large because pilgrims do not forget to fulfil this formality. Women who enter the chapel which encloses the saint's tomb, fall into a fainting fit, something which happens to women at the tomb of Mas'ud Ghazi too, and for which Afsos offers a detailed explanation which can be seen in my *Memoir*.[22]

The information Mrs Hassan Ali gives about Shaikh Saddou (or Saddo) are in conformity with many points which I have borrowed from Roebuck[23] and mentioned in my *Memoir*.[24] But according to Mrs Hassan Ali it should appear that people do not consider him a saint but rather a bad spirit who seizes men and throws them into melancholy, hypochondria, etc. In order to be cured, the person affected by this disease distributes sweets to the poor and in addition, if he can afford it, the sacrifice of a black goat. Blindness, they say, is produced by the same cause; it is cured by roasting the liver of a kid and eating it immediately.

The twenty-sixth letter deals with the superstitions of the natives and the twenty-seventh, the last in the second and final volume, contains a detailed notice on Meer Haji Shah, the author's venerated father-in-law.

I should finish this article with an observation to which I desire Mrs Hassan Ali should pay attention in case she wants to render a second edition to this interesting and important book. Her orthography of the transcription of Hindustani words is defective, so much so that it is difficult at times to recognize them. I have noticed, especially, that in many of the words which have an *r*, *re*, she omits this letter, while in other cases she often adds this, where it is not necessary. One reads, for example *sota badhaa* for *sota* or *sonta bardar*, a sort of masseur: *mortem*, for *matem* or *matam*, mourning: Ayashur for A'isha, Muhammad's wife: *maivour* for *maivoua* [*sic*], *mewa*, fruit, etc. It is regrettable that this work so interesting and so useful should be marred by such errors which the author could have avoided easily by submitting it to an Orientalist.

I should say again that Mrs Hassan Ali has taken the liberty of introducing an innovation which I am far from disapproving; this is anglicizing certain Hindustani words, such as *salaaming*, wishing *salaam*; *purdahed*, behind the *pardah*, etc. English gazettes of India are full of Hindustani words which do not have proper equivalents in English, but I have rarely noticed such words dressed up in English.

The publication of another book on Indian Muslims, one translated by Dr Herklots, of a Hindustani manuscript entitled *Qanun-i islam dakhan, Law of Islam in the Deccan*, is announced in London. As soon as it is published I shall make it known to the readers of *Journal asiatique* all the more willingly as I had the honour to have the editor present at my lectures and I am convinced that one will find in this work valuable documents and entirely new details on Muslims in that part of India.

NOTES

1. As Mrs Hassan Ali mentions the death of her father-in-law but does not say that she has lost her husband, I believe that the latter is still alive.

2. *Nouveau journal asiatique*, vol. VIII, p. 90 and p. 12 of the independent publication.

3. Ibid., pp. 167, 36.

4. Ibid., p. 165, 34.

5. Mr de Sacy, *Chrestomathie arabe*, vol. 1, p. 49, new ed.

6. See my Memoir, Nouveau Journal asiatique, vol. VIII, p. 214; p. 83 of the independent publication.

7. *Nouveau journal asiatique*, vol. VIII, p. 167, p. 36 of the independent publication.

8. Vol. 1, p. 159.

9. Ibid., p. 165 sqq.

10. *Nouveau journal asiatique*, vol. VIII, 199; p. 69 of the special ed.

11. Elephants are so common in Lucknow that Mrs Hassan Ali knew a beggar who went about begging sitting on this animal. See vol. I, p. 276.

12. *Nouveau journal asiatique*, vol. VIII, p. 219; p. 89 of the special edition.

13. See III Kings, xvii, 1 and xviii, 1 and 43.

14. IV Kings, ii, 14.

15. French ed., p. 145.

16. See the analysis I have given in the *Bulletin des sciences historiques* (viith section of *Bulletin universel*). vol. XVIII, p. 292.

17. Page 139 *sqq*, Calcutta ed.

18. In the same way almonds and roasted beans are sold in the streets of Marseilles.

19. Muslims have a great respect for the descendants of Muhammad and have a very high idea about their spiritual prerogative. Wali says in one of his *ghazals*, 'O Syed don't be afraid of the day of judgment, because the family of the Prophet has nothing to dread.'

Na dar rozi mahshar sati saiyidi
Ki āli nabi par na awega al.

20. *Nouveau journal asiatique*, vol. VIII, p. 322; p. 104 of the special edition.

21. The same who under the name Maulawi have recently raised a revolt which has been mentioned in *Asiatic Journal* N.S., vol. VIII, *As. int.* p. 14. In the same number it has been said that they are Sunni and that their leader is still alive.

22. *Nouveau journal asiatique*, vol. VIII, p. 210; p. 79 of the special edition.

23. Ibid., p. 182; p. 49 of the special ed.

24. *Persian and Hindi Proverbs*, vol. II, p. 26.

3
Islamic Practice:
Reflections on Ja'far Sharif's
*Qanoon-e Islam**

I

The work I am going to review is undoubtedly one of the most important books ever published on the religion of Muhammad; the huge material which has been gathered in 592 large pages in octavo would have easily filled several volumes. As a result, a satisfactory analysis of the book should have been possible only in a series of articles, but on account of the limited space, I have chosen those sections which appeared to me most interesting and I wish to talk about the religion of Indian Muslims, a subject which occupied me in my special *Memoir*.[1]

The author of this book, Ja'far Sharif, nicknamed Lala Mian, is a Sunni Muslim, and a native of Ellora

---

* The full title of the work is *Qanoon-e Islam, or the Customs of the Moosulmans of India: comprising a full and exact Account of their various Rites and Ceremonies, from the moment of Birth till the hour of Death: by Jaffar Shurreef (a native of the Deccan), composed under the direction of, and translated by G. A. Herklots, M.D.* London 1832.

in the old kingdom of Golcunda. He tells us that educated people have regretted that there is no complete
book on Islam as practised in the Deccan, and Doctor
Herklots succeeded in persuading him to write a
treatise which would trace the duties of Muslims from
birth to death, and their social and religious customs.
Following faithfully the plan he had adopted, Ja'far
Sharif describes in the first chapter the rites and
ceremonies observed by Muslim women during
pregnancy and confinement; he also describes the
ceremonies which take place on a child's birth, and
incidentally the four classes or races of Indian Muslims, the Syed, the Shaikh or Mashaikh,[2] the Mughals
and the Pathans or Afghans. The Syeds call themselves Hasani, Husaini or Alvi according to whether
they are descendants of Hasan, Husain, or Ali's other
children. A Syed's wife is called Syedani. The name
Mirza (son of a Mir, that is of a Syed) is given to the
son of Syedani and a Mughal. The Shaikhs are subdivided into three groups; the Quraishi, who belong
to Muhammad's tribe, the Siddiqui, descendants of
Abu Bakr, nicknamed Siddique, the truthful, and
Farooqui, descendants of Umar, nicknamed Farooque,
the discriminator. There are two types of Moghuls ,
the Iranis of Persian origin, are Shia, and the Turanis,
or Tartars are Sunnis; finally the Pathans, who claim
to be descendants of the patriarch Jacob, subdivided
as Yusuf zai, descendants of Jacob and the Lodi, descendants of Loth. Then there are those who call themselves Durrani, on account of the pearl (*dur*) they
wear in the ears, and *gair-Mahdi* (non-Mahdi) who do
not believe in the continued existence of the last
Imam and in his appearance along with Jesus Christ
at the end of time. They are also called Abdali and

daire-wale. In addition to these four principal classes, there is another called the Nawait [*sic*] *nawa-aé-thé*, or the new comers.

While talking of the Muslims races of India, he does not forget to tell us about the two sects in which the Muslims of the country are divided, that is, Imami and Sunni. The author, who belongs to the first sect, always calls his adversaries Shias, a word which means sectarian; and he claims that in India they are less in number than the Sunnis; though one may agree with Mrs. Hassan Ali[3] that they are more or less equal in number or, what is more probable, is that the Shias are in a majority.[4] Ja'far is, however, tolerant as Muslims in India generally are and he finds very ridiculous the excessive zeal which leads some Imamis to refuse to pronounce the number 4 in arithmetical calculation and not to use the cot with four legs, *charpai*, and other puerile practices on account of contempt for the number 4, which was the number of Caliphs which the Sunnis acknowledge as legitimate. He objects very strongly to what some Imamis do on the festival of Ghadir.[5] According to him:

> They shape into three hollow figures, pastry made of leavened flour and fill them with honey, and after cursing these figures which are supposed to represent Abu Bakr, Umar and Usman, they pierce a knife into each one of them; then they suck the honey which flows out of it and eat the pastry, as if they wanted to make people believe that they are thus drinking the blood and eating the flesh of their enemies.

In chapters II to XII Ja'far and his translator, Mr. Herklots, describe at length the numerous ceremonies

which are observed from the birth of a child to his
first birth anniversary, which is called *sal girah* or
baras-ganth, or the knot of the year, because on this
day they tie a knot of a cord ad hoc, in order to mark
the person's age. When the boy is four years old, they
make him pronounce *bism-illah* (in the name of God)
and between the ages of seven and fourteen he is cir-
cumcised. And at the time of the boy's or the girl's
puberty, there are new ceremonies which have been
described faithfully by the two authors.

Chapters XII and XIII describe the fundamental
principles of Islamism, which have been the theme of
a very large number of books, but it was considered
necessary to include it here to make the treatise complete. About the pilgrimage to Medina, the author
makes an observation which highlights the difference
between the semi-idolatrous Indian Muslims and the
true Muslims of Arabia and other countries who follow the Quran . The veneration of Indian Muslims
for the saints is as great as that for God, and that is
why those who visit the Prophet's tomb make it a
point to prostrate themselves before these sacred
relics. But, says the author, the attendants of the tomb
prevent them from doing so and remark that *sijdah*
(adoration by prostration) is a mark of respect
reserved for God alone, and performing it for a
created being is a sin. Ja'far, of course, disapproves
of the practice of Indian Muslims prostrating themselves even before a *tabut* or *ta'zia* (representation of
the tombs of Husain and Hasan) and the banners of
processions.

I do not speak of chapter IV which deals exclusively with the religious and non-religious ceremonies observed on the occasion of marriage. One goes through

them with interest even as one has read the descriptions given by Mackenzie[6] and Mrs Hassan Ali,[7] because the information provided here is extremely detailed and exact and useful Hindustani words accompany the explanation. Although the Muslim religion permits the remarriage of widows, four months and ten days after the death of her husband, such a marriage is rare; it is in this as well as in other matters that Hindu prejudices have subjugated the Muslims.

Chapter XV which comprises eighty-one pages deals with the festival of Moharram, celebrated mainly as we know in memory of the martyrdom of Husain. Even though he is a Sunni, the author mentions with horror the name of dirty (*palid*) Yezid, who got Hasan poisoned and Husain massacred. According to him, it was not Hasan's wife who poisoned him but Joada, a woman seduced by Marwan, Yezid's minister. Jafar narrates at length the touching story of Muslim and his two sons, a narrative, which serves as a prologue to that of the sad end of Husain and his seventy-two companions.

In India they have a special building for the celebration of the festival of Moharram. This building is called Imambara (House of Imam) in the North and East but in the Deccan it is called *ashur-khana* (house of ten days), *ta'zia khana* (house of sorrow) or just *astana* (threshold). They put here the *tabut*[8] or replica of the tombs of Hasan and Husain; the *shahnashin*, a kind of small portable chapel, whose sketch one can find in the Herklots' book;[9] the *alams* or banner which are carried in the Moharram processions; the figures of the *buraq*, the animal with the body of a horse and head of a woman, with the wings and tail of a peacock, the animal which carried Muhammad to Heaven. All these objects are surrounded by a *tatti* or frame made of mica which shines and reflects the light. There are also figures made of wax and other materials, of men and animals, plants and fruits. In front of the *ashur-khana* is dug annually a circular ditch called *alawa* (or more correctly, *alao*), that is, fire of joy, since every evening during the period of the festival a fire is lighted. The common people, young and old, amuse themselves, running and playing and crying, *Ia Ali* (O Ali), *Shah Hasan* (King Hasan), *Shah Husain* (King Husain) etc. There are also those who cross the flame or jump on the fire to scatter it to a distance. Sometimes they dig an *alawa* without an *ashur-khana*, and during the ten nights of Moharram, women recite before the fire *marsias* (elegy in honour of Husain) and from time to time beat their breast. Some of them content themselves by placing a lighted lamp on a reversed wooden grinder or an earthen vase. On the first, third and fourth days of the lunar month, they decorate the *ashur-khana* with carpets, transparent screens, lamps, chandeliers, candles, ostrich eggs, paper flowers, etc. In each of these buildings they plant into the earth five or six banners generally called *alam* or *chhada*. These banners or flags are in so many different shapes that it will take too long to describe them here. The piece of cloth attached to the pike is generally triangular; it is red or green, red for Husain and green for Hasan. One can also see the figure of *zulfiqar* (Ali's sword) or the *qadm-i rasool* (footprint of Muhammad). Before these standards are placed lamps, *morchchals* (fans), *oud-soz* (cassoulets), etc. Incense is burnt in front of the *alams* and it is proper to remark that incense is used in the Indian

Muslim rituals for God as well as for saints, images or relics etc. During the day they read the Quran in the *ashur-khana* and at night, boys sing *marsias* surrounded by *faqirs* and a large number of other listeners.

On the afternoon of the seventh day of the festival, in the Deccan they take out the banners of Qasim, nephew and son-in-law of Husain; this banner is distinguished from others as it has small parasols of gold or silver on the top. It is carried by men on horses or on foot and is accompanied by troupes of musicians and bayadères who sing *marsias* and beat their breasts crying, *Doulha, Doulha* (bridegroom) alluding to Qasim's marriage.[10] They honour this banner by fanning it with a *morchhal* and keep incense burning so long as the procession lasts. When they return to the *ashur-khana* they put the banner on a bench ; then they offer *fatiha* over two or three jugs of sherbet called *ran ki* [sic] *sherbet* (sherbet of the war) which they distribute among those present.

On this day they carry a pike (*naiza*) on which a lemon is impaled and from it they suspend a turban, a shawl, a bow and two swords. The lemon represents Husain's head, which was on Yezid's order carried from town to town on the point of a lance ; this procession passes in the same manner as the earlier one. In the evening *nal saheb* is exhibited; it is a replica of Husain's horseshoe, made of iron or another metal, ordinarily of wood or even of paper coated with sandal powder; this figure is of course larger than the ordinary horse shoe. The person who carries the *nal saheb* runs like a man possessed and knocks down anyone who comes in his way—old men, women, children. He wants to imitate by his fast and light movements, Husain's horse. He is accompanied by young men holding *morchal* (fans), the *aftab-gir* (parasol) or simply swords or even sticks, and these persons generally participate in this ceremony only to fulfil a vow taken by their mothers. If a woman has promised that if she gets a child from Heaven she will make her child run with the *nal saheb*, she cannot afford to back out of the promise. Everybody is free to organize a procession and that is why one sees banners converging in the streets from different points, and it is often the custom to touch a banner with others, offer *fatiha*,[11] burn incense and then pass on. They also take out on this day, figures of the *buraq*, the animal which we have already mentioned. It is made of wood, is two or three feet high, and is painted and decorated elegantly.

On the tenth day they take out all the flags, except Qasim's. Men carry the *tabut* on their shoulders; they are surrounded by *faqirs* of all types and other people carrying torches and flames. Ordinary people take out these processions in the evening, while distinguished persons do so at midnight. The streets are illuminated and everywhere there are magic lanterns and china shades representing the battles, and this attracts a large number of people. Throughout the night the whole city is alive and the noise and tumult does not cease even for a while. In place of the *tabut* or the *ta'zia*, some Muslims make use of a *shah-nashin* (room of honour), also called *dad mahal* (palace of justice), a monument which is like the former but with the difference that it is made in the form of a palace, and at each angle there is a sort of lamp, with a transparent screen with figures of animals, which turn more or less like the lampshades one sees in some shops in Paris.[12]

Faqirs play a great role in Moharram processions. Most of them disguise themselves grotesquely. The Jalalia are found in great numbers; some of them put on very big turbans made of straw; others have around their necks rosaries with the beads of fruits of all kinds; they have half their faces painted black and they carry horrible-looking dolls. Others disguise themselves as *biaj khor* (usurers), *murda farosh* (gravediggers), *baqqal* (fruit-sellers), *sharabi* (drunkards) *galiz shah* (cesspool cleaners). Some of them are called *majnun* (mad). Some represent Qais the lover of Laila or Laili; they have their heads covered with bonnets made of sweet bread with a long paper tail which sweeps the ground; sometimes they carry lanterns on their heads; they go to the *ashur-khana*, where they dance in a circle more or less like the *maulwis* in Constantinople. Other *faqirs* represent Laila; they wear three-cornered paper caps; their dress is similar to the disguise which prevented Abbé Scarron from the use of his limbs; to achieve this they rub soot all over the body and then roll on cotton. There are some who disguise themselves as negroes and negresses; others would like to be called *Haji Ahmaq* and *Be-waquf* (mad and idiot pilgrims); they have hemp wigs, beards that reach down to the navel and false monstrous bellies; they indulge in a thousand clownings to make the people around them laugh. There are also those who disguise themselves as old men and women; the latter cover their faces with masks, with a big *nath* or ring suspended through the nose. Others are disguised as doctors and they follow a horse carrying a large number of small bags full of simples [herbal remedies].

There are those who call themselves Mughals; they have rosaries and carry sticks and give themselves ridiculous names, like *Gajar Baig* (Lord Carrot), *Shalgam Baig* (Lord Turnip), *Mirchi Baig* (Lord Pepper), etc. They talk among themselves in a kind of *argot*. Others disguise themselves as nabobs, riding a horse and followed by a number of domestics. What is remarkable is that Muslim *faqirs* disguise themselves as Hindu *yogis*; like the latter they play on different musical instruments and sing hymns adapted to the particular situation. Others disguise themselves as *Nanak panthis*, followers of Nanak; they carry clubs in each hand and strike one against the other. There are also attractve young boys dressed in women's apparel. Finally, there are *faqirs* who lie on the ground in the *ashur-khana* with blood-stained swords by their sides, hiding their head, making it appear that there is a head without a body or body without a head. They wish to represent with this mimicry the martyrs of Karbala. There are also those who disguise themselves as bears (*Rich Shah*), tigers (*Bagh Shah*), or camels (*Unt Shah*). The last masquerade is managed by a man moving like a camel as they do in our theatres. Up to now I have mentioned only comic disguises; but those I am going to talk about now cross the limits of pleasantry. There are *faqirs* who are called *Qazi Lain* (cursed judge), *Qazi be-din* (impious judge), who are dressed like Muslim judges and parody religious discourses in a manner which is really scandalous; for example, 'He who prays, fasts, etc., shall be ELEVATED to hell; he who commits adultery, practises usury and allows himself to be corrupted by presents shall be PRECIPITATED to Heaven.' Others parody a procession; they carry a pole on which they put dirt and cover it with twelve pieces of gold or silver cloth; if some curious person desires to see these fake relics they seriously uncover the the twelve pieces of cloth one after the other and burst into laughter at his disappointment. They call themselves *faqirs* of *St Honorable Husain.*

It would be much too long to report all the follies which mark this exceptional festival; what has been said is more than sufficient to show how it has degenerated into a veritable masquerade which makes it resemble the Holi of the Hindus.

A large number of Muslims and even Hindus make a vow to execute particular practices at the time of Moharram, if they receive from Heaven the boon they wish. The most common in the Deccan is the bringing out of an anchor (*langar nikalna*), and the supposed anchor is nothing but a silver or iron chain covered with garlands of flowers which are worn round the loins and which trail on the ground. Others take a vow to sweep with their hair the sides of the *alawa* or give a bath to their heads in the fire of the *alawa*, or not to break the fast observed on the occasion except with food received from begging, etc. Others promise to reach the place where the *tabut* is buried by rolling on the ground, on the tenth day called *shahadat ka roz*, the day of martyrdom, or at any rate to cover part of the distance in such a manner. To bring good luck to their children, several people on this very day fix on their ears, under one of the *tabuts* carried in a procession, gold ear-rings, and if it is a girl, a ring (*bulāq*) through the nose. On the night of the ninth day, a grand nocturnal procession of Hindus and Muslims which could be called a 'general procession' (*Moharram ki shab gasht*) gathers in the town from ten miles around the city. On this day they first offer in the name of Husain, *fatiha* over pilau, sherbet, etc., and then they take out the banners, the *tabut*, the *buraq*, etc. Ordinary people holding torches accompany them and also people carrying incense and funny *faqirs*, as already mentioned. The procession is preceded by a troop of musicians and a band of *bayadères* who recite *marsias*. They first make three rounds of the *alawa* and then they march in the streets and bazars and at dawn they return to the *ashur-khana* from where they had started. On the last day of the festival[13] they take out a procession similar to that of the previous day, but since it is taken out in the day from nine in the morning to three in the afternoon, there are no torches or flames. It is on this day that they go to the plain of Karbala (*Karbala ka maidan*), which means simply a plain situated near the sea, a river or a pond; it is on this day that they carry the *tabut* for burial. On the way they throw cowries on the banners; women and children gather them enthusiastically because they are considered holy. The plain of Karbala offers the spectacle of a veritable fair; on all sides there are jugglers and buffoons; there are stalls where fruits and betel leaves are sold and *abdarkhana*, a kind of café, where one can get various kinds of refreshments; finally men carrying water-skins distribute water to those who want it. When the procession reaches this place, *fatiha* for Husain is read over pilau, sherbet, cakes and sweets, which are then distributed in part among those present; then they take out from the reliquary the two reproductions of the tombs of Husain and Hasan, and lower them in the water. They bathe only the *alam*, taking care to remove whatever else could be soiled by water. Sometimes,

besides the reliquary of Husain they have another
kind of *tabut* called *ran ka dola* (the palanquin of the
battle.) It is made of bamboo and covered with white
cloth; it has seventy-two compartments, supposed to
represent the chests in which the heads of the martyrs
of Karbala were put.

Chapters XVI and XVII deal with the ceremonies
(fêtes) held in Safar and Rabi'ul auwwal, the second
and third months of the lunar year of Muslims. Commemoration of the death of Muhammad takes place
on the 12th of Rabi'ul auwwal; and taking this fact
into account, they call it *bara wafat*, the 12 (of the)
dead, and in this regard, I should make the observation that the explanation offered by the Hindustani
writer Jawan is, I believe, not correct—an explanation
I have given in my *Memoir* on the Muslim religion in
India.[14] During the first twelve days of the month, the
Quran is recited every morning and evening, in
mosques or in private houses and alms are distributed
after the *fatiha* is recited and incense is burnt in the
name of the Prophet. The story of the death of
Muhammad (*wafat nama*) is also recited in Hindustani;
it is useful to point out that, generally speaking, in
India hymns, sermons and even supererogatory
prayers are recited in Hindustani. Only the obligatory
prayers are offered in Arabic and some specific hymns
in Persian.

On the eleventh day they go in a procession to a
qadm-i rasul, that is to say, the Prophet's footprint.
These are imprints on stone and are not very rare in
India.[15] They carry on a stretcher one or several dishes
of sandal powder and perfume called *argaja*, and
cover them with a sheet of flowers (*phool ka chadar*).
When they reach the *qadam-i rasul*, they dip a finger

in the dish of perfume and apply a little of it to the
sacred footprint. The twelfth day is called *urs*; this is
the name they also give to most of the festivals of
Muslim saints in India. This is an Arabic word which
signifies marriage, in the same way as does the Persian synonym *shadi*, and the Hindustani synonym
byah.

II

The chapters that follow those which I have already
introduced, that is, the eighteenth, nineteenth and
twentieth, describe the festivals of three great saints
of the Muslim Deccan: Pir-i Dastgir (that is, the
Protector Saint), Shah Madar and Qadir Wali Saheb.
Pir-i Dastgir is the honorific of Syed Abdul Qadir
Jilani, to whom I have dedicated a chapter in my
Memoir on the Muslim religion in India. His fête is
held on 11th Rabi'us Sani, the fourth lunar month of
the Muslims. If one were to believe the author of
Qanoon-i Islam, Abdul Qadir is the same person as
Miran Mohiuddin, whose fête according to the author
of *Bara Masa* falls on the second of the same month.[16]
Since the term Mohiuddin is an honorific which
means one who revives religion, it is quite possible
that this title has been given to the same person; but
I doubt if the Mohiuddin, about whom I talked in my
Memoir, is the Abdul Qadir, whose grave is at Baghdad, while according to the description given by
Jawan, Mohiuddin's grave is in India. What appears
to be nearer the truth is that Miranji, also called
Sheikh Saddou, is a different person and does not
deserve to be counted among the saints of Muslim
India; Mrs Hassan Ali and Mr Herklots agree on this
point. Sheikh Saddou is considered a bad spirit, who

can possess any person he chooses; however, the low-born, and specially their women, whom Ja'far considers irreligious, have made a cult of him and his tomb attracts a large number of people.

The article on Shah Madar gives some new information; the saint is called Zindah Shah Madar, that is to say, the living Shah Madar, because, while the most reasonable among his devotees think that he lived for 395 years, others believe that he is still alive. Jafar claims that the saint never married and that he had no knowledge of women; thus all the 1,442 sons that another writer[17] has given him are his spiritual children as I also had thought.[18] His fête takes place on the 16th and 17th of Jamadiul auwwal. It comprises several distinct ceremonies: on the 16th, they take out the procession of the sandal, which has already been referred to in the note on the death anniversary of Muhammad; on the 17th they prepare pilau and other food, on which they place seventeen lamps; after the fatiha is recited the ceremony of the *baddhi* (a kind of collar) takes place. In this ceremony the collar is put around the children's necks, in honour of this saint. This practice is generally observed by parents who have taken a vow that if a male or female child of their choice, is born to them, it will be dedicated to Madar. The collar, ordinarily of gold or silver, or even of flowers is a sign of this consecration. I have said elsewhere,[19] that the *faqirs* belonging to Madar's Order, called Madari or Tabqati, cross bare-footed, the fire, which is especially lit for this purpose. Ja'far informs us that people enthusiastically prepare piles of live charcoal for these *faqirs*; they call this ceremony Dhammal Koudana (to jump into fire.) Presents are made to *faqirs*, and after the

ceremony, their feet are washed with milk and san-
dalwood and they say that the fire does not burn
even a hair on the *faqir's* body. On the 17th of this
month, some Muslims slaughter a black cow in the
name of Madar, either in the *ashur khana* or *astana*, or
in their own houses and distribute the meat among
faqirs.

Qadir Wali Saheb, whose fête is celebrated on the
11th of Jamadius sani is, according to Mr Herklots,
the same person whom I have called Mo'inuddin
Chishti. But I do not agree with Mr Herklots:
Mo'inuddin is buried in Ajmer, and Qadir Wali at
Nagore, near Negapatam, more than 800 miles away
from Ajmer. There is nothing surprising if Jafar, an
inhabitant of the Deccan, does not talk of Mo'inuddin,
who is famous in the north and east of India, and if
Qadir Wali is not mentioned at all by the writers of
the north and east whom I have consulted. There are
saints who are very famous in one province of India
but unknown in other parts of the country.

Ja'far gives a very detailed description of the
ridiculous ceremonies which mark the saint's fêtes but
unfortunately he does not give any biographical facts
about him, except the three absurd miracles which
Herklots has been kind enough to translate in full in
order to keep the original text intact. This saint is
honoured particularly by sailors; the *nakhuda*[20] and
the *laskar*, or to be more precise, the *khalasi* (sailors)
often take a vow in moments of distress to offer in
his honour a *fatiha* or an *ex-voto.*

Chapter XXI describes, among other things, Rajab
Salar (not Salars), whose tomb is situated in Baraich
[*sic*] [21] in the kingdom of Oudh. This martyr saint is
invoked against diseases of the leg; those who are

cured through his intercession, offer during his festival, figures of horses made of pastry, called *khule
ghore* (free horses).

It appears that in the Deccan a special fête is held
in memory of *mi'raj*, or Muhammad's ascension. It
takes place on the 15th, 16th or 27th of Rajab, depending on the locality where it is celebrated: but Ja'far
adds that only the pious and the educated participate
in this fête.

The Commemoration of the Dead, Shab-i Barat (the
Night of Deliverance) already described in chapter
XXII, is one of the two Muslim festivals which are
preceded by a vigil; the other is Baqr-Id, which Christian writers compare to Pentecost. Among the offerings which are made for the peace of the soul of the
dead, I should mention elephants and lamps made of
clay; they are put on a bench and around them rice,
sugar, dates, almonds and other fruits are arranged.
On the preceding day there are illuminations and Mr
Herklots remarks that at this festival there are a lot
of fireworks and people enjoy aiming one piece
against the other, in spite of the regrettable accidents
which take place quite often. On this occasion friends
offer fireworks as presents.

I shall not say anything about chapters XXIII and
XXIV which are about the month of Ramazan, and
about the festival which is celebrated at its end. In
the Deccan, the month of Shauwwal is called the
empty month (*khali mahina*) since there is no festival
in this month. The festival which we have just mentioned, the festival which is called the Easter of Muslims, is supposed to belong to the earlier month; in
the north and east of India, on the contrary, the name
khali is given to the following month (Zi-qada) be-

cause it is really this month in which there is no festival; this is not the case in the Deccan where on the
16th the fête of Banda Nawaz, (patron, literally,
cherisher of servants), the honorific of Gesu diraz,[22]
is held. This fête is more solemn than that of Qadir
Wali's. This is especially so at Kalbargah [sic] where
his tomb is situated and where his devotees go to
celebrate it. There the ceremony of the sandal is observed on the 16th and the *urs* on the 17th. According
to a vision this scholarly saint had, the Muslims of
central India believe that those who for valid reasons
cannot make a pilgrimage to Mecca but who have
visited even once, the tomb of Banda Nawaz, will
achieve as much merit as they would have if they had
visited the place prescribed by the Quran.

Baqr-Id is the theme of the following chapter. This
and the festival which is held at the end of the month
of fasting, of Ramazan, are the only two real festivals
of the Muslims; the others are accessory, supererogatory solemnities more or less tainted with
idolatry; they vary from place to place; each province
of India, even each town, each village has its own one
or several patrons, for whom a fête is celebrated. Thus
Ja'far tells us that in Hyderabad, for example, people
extend special honour to one Maula Ali, whose fête
is held on the 16th and 17th of the month of Rajab.
'In order to have a correct idea,' says Ja'far, 'it is
necessary to witness the crowd which is attracted
there and the resulting noise.'

Our author comes back in chapter XXVIII to vows
and oblations (*nazr o nayaz*) which have been incidentally mentioned in the context of Moharram and other
festivals. One of these oblations is in the name of
Khwaja Khizr and this consists in launching in the

sea, or a river, a kind of boat of which a sketch has been given by Herklots in plate 4. Some of these are in the form of a peacock and are called *mohar pankhi* (wings of a peacock), while others are simpler in design and are called *lachka* (boat) or *bera* (raft). They are very artistically made of bamboos and are covered with coloured shining paper. Above it is placed a kind of *tabut* in which they put lighted candles; the *tabut* is made of mica and is called *kanwal* (lotus) because it is decorated with flowers of this plant. With great pomp these boats are carried to a riverbank or seashore; they are there launched on the waves amidst fireworks and the sound of music; and immediately after, children and even grown-up men plunge into the water and swim after them and seize them avidly. Other persons content themselves with putting on the bank clay or copper lamps on which they put cowries or paisa.[23] Children seize the coins; however, if it is a copper vase the owner takes it back. Besides, a large number of clay lamps are floated in the river one after the other; those who possess considerable wealth float a boat large enough to carry several hundred people. This boat is lighted and decorated with *mohar-pankhi* and fireworks are set off as the boat advances to the middle of the river. Poor people launch clay pots in which they put packets of betel- leaves, some *suparis* (areca nuts), rough sugar cake wrapped in banana leaves and a lamp with ghee (refined butter). They give some cowries to the *mullah* for the *fatiha* to be recited at this exceptional expression of piety. They come back with a *lota*[24] of water collected at the place where the ceremony is performed; they break their fast with a gulp of this water and then eat their meal.

It is no use citing other saintly persons to whom
oblations are made with a view to fulfilling a vow;
however, there is one whose name I must mention,
since one may be astonished to see his name in the
diptyque of Indian Muslims; this saint is Sikandar,
Alexander the Great, to whom good Muslims vow to
offer clay horses if, through his intercession, their
desires are fulfilled; they make these figures of horses
along with riders and take them ceremoniously to a
certain place where hundreds of such figures are piled
up; recitation of the *fatiha* precedes this ceremony.
Hindus share the Muslim veneration for this hero of
Macedonia.

The next chapter, one of the most interesting in the
book, describes the different classes of religious Muslims, or *faqirs*, who can be seen in India; one discovers
in this chapter that in India there are religious women
(*religieuses*) as well as religious men (*religieux*). When
a man or a woman wants to become a novice of a
silsila (order), in other words, a *murid* (disciple), he or
she chooses the *pir* or *murshid* (superior of an order);
the latter, after having performed *ouzou* [*sic*] (*wazu*)
(ablutions), takes the former's right hand and holds
it for a while; he does not hold a woman's. During
this time the disciple recites a sort of confiteor, then
the director introduces him to his spiritual genealogy,
which always goes back to the Prophet, and then asks
if the novice accepts the authenticity of this genealogy; on his affirmative response, the *murshid* lets go
the disciple's hand and performs certain other
ceremonies, after which the newly elect is declared a
novice; he receives a copy of his spiritual master's
shijra (genealogical table), which becomes his also.
Some fanatics consider this *shijra* more sacred than

the Quran itself ; they put them in amulets which they wear on the arm or round the neck; and when they die they are buried with this genealogy put on their breast. Even after the novice becomes a profès, there are a series of ceremonies to be performed before the elect puts on a new dress and receives from the *pir*, who has admitted him to the community (*jam Allah*, assembly of God), a name which always ends with the word *Shah* (king) in order to indicate that he has now become the master of his own passions.

There are four principal *pirs* and fourteen spiritual families to whom all the orders of *faqir* are related. These four *pirs* are Ali, who invested the spiritual Khalifat to Khwaja Hasan Basri, who transmitted his rank to *pirs* Khwaja Habib 'Ajmi and Abdul Waheed bin Zaid Kufi. From the third *pir* descend nine spiritual families; and from the fourth, five families, equally spiritual, trace their origin; one finds a precise list in *Qanoon-i Islam*.[25] Then we have a list of religious orders which are most widely spread in India; they are ten; but it is not possible for me to tell about all of them in this brief article.

I would limit myself to saying that the *madaria*, who consider Madar as their founder, are jugglers rather than true *faqirs*; they rear tigers and other animals to show them to the curious; they make bears and monkeys dance in markets and execute a thousand skilful feats; they always wear black clothes; their *pagri* (turban), *jama* (robe), *dopatta* (shawl), *loung* [*sic*] a kind of underpants; everything they wear is black. The founder of the *malang* was a disciple of Madar. Apparently they are reformed *madaria*. They go about naked, or at best they have only a short loin cloth as dress; their customs are similar to those of Hindu *faqirs*, particularly to the group called Gosain.

The *rafai*, also called *gurz-mar* (that is, strikers of clubs, since they strike their chest with a club) consider Syed Ahmad Kabir as their founder. Mrs Hassan Ali has talked about them in her *Observations on the Musulmauns of India*, and sometimes they are mentioned in the Gazettes of India. They practise revolting austerities, for example, striking themselves with swords, cutting off their tongues or burning them with hot irons, putting living scorpions in their mouths; they often go to shops armed with their *gurz* or clubs, and if the shopkeeper does not give them alms according to what they presume to be his means, they brandish their weapons in order to frighten him. Out of fear, the shopkeeper gives them some *poisa*, but sometimes they throw it away contemptuously since they do not want people to accuse them of extorting alms. The Jalalia have Syed Jalaluddin Bokhari as their founder. When they are admitted to the order, a spot on their left arm is cauterised with a burning piece of cloth; the scar is their mark of distinction.

The Sohagia, who draw their name from Musa Sohag, are distinguished by the female dress they wear. Generally, they wear on their wrists *chori* and *bangri* (a kind of bracelet made of particles of glass) and receive alms especially from the *kanchani* (bayadères); if anybody refuses to give them alms they break the bracelet and eat the particles. These *faqirs* play on different musical instruments, dance, and sing, not only among themselves but also in front of the curious who desire to listen.

All these *faqirs* are either *ba-shar'* (with the law) or *be-shar'* (without the law). The former practise the external precepts of the Muslim religion; the latter, who are larger in number, do not take these precepts into account at all; they intoxicate themselves with *bhang*,[26] opium, wine, beer, and other fermented drinks; they do not fast, do not offer the prescribed prayer and some of them indulge their dissolute passions without any check. The *be-shar' faqirs* do not marry; they live like vagabonds or tramps; they sleep wherever they find themselves; if anybody gives them something to eat, they eat it; if nobody gives, they go away. Muslims consider them generally to be saints and believe that they do not submit themselves to the duties which an ordinary believer is obliged to, because they are absorbed in their meditation on interior doctrines. Some of them assemble at real convents called *takia*.

I shall say nothing about chapters XXIX–XXXVII, which deal exclusively with exorcism, amulets, charms and all that is related to the fake art of magic, although, in these chapters, there is a lot of extremely interesting material, which may not be found elsewhere. The three final chapters of *Qanoon-i Islam* deal with ceremonies which precede and follow the death of a person as also with all the practices of piety, related to the dead. In India, an ordinary tomb is generally made of clay, large on the side of the feet and narrow at the other extreme; on it a channel is made, which is filled with water. On the side where the head is placed, they put an inverted pot and quite close to it, a pomegranate tree. Besides the funeral service which is held on the day of burial itself, they offer in India new and special prayers peculiar to the country, on the third, tenth, twentieth, thirtieth and fortieth days after the death and at the end of the third, sixth and twelfth months.

It remains to talk about the appendix which would by itself have made an important book. Among other things, one finds there an exact and detailed list of Hindustani names of male and female garments, jewels and all the *mundus muliebris*. The notice on musical instruments known in India, with their illustrations, and the notice on games are not less interesting. The glossary comprises more than 150 words, the precise significance of which is explained and developed at length, sometimes in several pages. The article dedicated to the explanation of the word *dargah*, which means tomb, contains a description of the famous *dargah* of the Deccan, in the village of Cuddry near Mangalore. It is built in the middle of a very wide cave, but the entry, six feet above the ground, is hardly large enough for a person to crawl inside. The *dargah* contains the relics of a *Shaikh* called Farid, who lived about a hundred years ago. He retired to this place, where he lived for twelve years; he lived there without drinking or eating anything, and without seeing anybody for forty days—an austerity which is called *chilla* (fortieth). After forty days he would come out of his hermitage, eat roots and wild fruits, drink water and meet those who came to see him. After four or five days he would return to his cavern from where he would come out again after forty days, and he followed this practice for twelve years, after which he disappeared, without anybody knowing what happened to him. Muslims visit this place on Fridays; they offer incense in his name, and recite the *fatiha* over food which they distribute among *faqirs*. During the reign of the unfortunate Sultan Tipu, the guardian of this *dargah* received from each of the boats, which entered the harbour of the town, from one to three rupees, depending on the number of masts; but this custom has ceased after the East India Company occupied this place.

The explanation of the word *khutba*, which is a sort of sermon or prone which is delivered on Fridays after the midday prayer, and which includes a prayer for the reigning king, gives the author the occasion to tell us that the sovereign for the whole of British India is actually Shah Alam's son, Akbar Shah II, the nominal King of Delhi in whose name the British govern and mint coins. Herklots tells us that in the Decccan *oud* (ud), does not mean wood of aloe as in Arabic and Persian and *agar* in Hindustani but benzoin, for which the Arabic expression is *bakhur jawi* or incense of Java. At the end of the book we find a satisfactory table of contents arranged in alphabetical order and plates which are carefully lithographed.

I should say a word on the role of Mr Herklots in the book which has been the subject of this article. First, his translation must have been extremely faithful, because he has done it under the author's eye and he must have consulted him whenever a passage appeared obscure to him or whenever an explanation was needed. To the role of an intelligent translator, Mr Herklots has added that of a scholar by providing explicative and philological notes and additions to the text, drawn especially from a very interesting work on the Muslims of India,[27] which an English lady, wife of a Muslim of Lucknow, published in England a few months before the publication of the book under review. These additions fill certain lacunae in the original book by expanding certain passages which, in the earlier book were rather too concise. The only comment that one can make on Mr Herklots is that he has depended for the transcription of Hindustani words on the vulgar rather than the normal pronunciation and has sometimes misspelt Arabic words. The last error comes from the fact that several Arabic letters are merged or confused in Hindustani; thus, the letters, *zad, ze, zal* and *zo'* are all pronounced as our *z*; the *sin, svad* and *se* as our *s* or *c*; the *he (bari)* and *he (chhoti)* as *h*; *khe, qaf* and *kaf* as *k*; the *te* and *to'e* as *t* and the *'ain* is not pronounced at all. It is impossible to give a correct orthography of Arabic words, if one followed Indian pronunciation and did not know the grammatical forms.

Here are a few words which have been transcribed wrongly; page xxiii of the appendices: *tawiz,* amulet, for *b'awiz* [sic]; page xxvii in the same part: *maniqa* (to embrace) for *m'aniqa*; page xcix. ibid.: *ta'waz*, recourse, for *ta'awwuz*; page c, ibid.: *tauwwaf*, walking around for *tawaf*; page 171, *camut* for *icamat* [sic] (kind of prayer), etc. The translation of a number of Arabic words is also not very exact; on page 163 the title *Rauzah ooch-chouhada (Rowzut-oosh-shohuda)* is translated as *The Book of Martyrs*, whereas the expression means the garden of the martyrs. Page 164, *zoul-janah* (and not *zool-junna*) is rendered as the winged wolf, while it means the wing of any animal; in this context it is the wing of a horse. In spite of these small errors, it is at any rate, and I repeat while rounding off, one of the most important books on Islam yet published; and one is very pleased with the painstaking translator for having it made known to intellectual Europe. The work will be consulted not only by Orientalists, but also by those who are not indifferent to the religions and philosophic doctrines of diverse communities.

NOTES

1. *Mémoire sur des particularités de la religion musulmane dans l'Inde, d'après les ouvrages hindoustanis, in-8,* 114 pages.

2. The Arabic plural is sometimes used for the singular. In Hindustani several Arabic irregular plurals are taken as singulars, such as, *nawab,* plural of *naib; oumara,* plural of *amir,* etc.

3. *Observations on the Musulmauns of India,* vol. I, p. 128.

4. *Asiatic Journal* N.S.; vol. VIII, p. 34 of *Asiatic Intelligence.*

5. For this festival see my *Mémoire sur la religion musulmane dans l'Inde,* p. 76.

6. *Transactions of the Royal Asiatic Society of Great Britain and Ireland,* vol. III, p. 170 seq.

7. *Observations on the Musulmauns of India,* vol. I, p. 350 et seq.

8. Last year I saw at Charonne, near Paris, two of these *tabut* or *ta'zia,* which had been made for a rich Muslim of Calcutta. It greatly resembled the one of which the design has been given by Mr Herklots on plate 1, fig. 1, but they were made with greater taste and were better ornamented and a peacock with the tail spread out was placed over the dome. One might compare them with a small gothic chapel with its turrets and diagonal ribs. Over the windows there were the following inscriptions:
   As salatu wassalam Ali Muhammad Mustafa o Ali Murtaza Fatimul Batul wal Amamin Hasan u Husain sallallahu alaihum o sallam.
   Greetings and peace on Muhammad Mustafa, on Ali the Elect, on Virgin Fatima, on the two Imams, Hasan and Husain. May God be propitious to them and accord His greetings.

9. Table 1, figure 2.

10. See my article on *'Observations on the Musulmauns of India',* by Mrs Hassan Ali, in the *Nouveau journal asiatique,* vol. IX, p. 542.

11. That is to say, the first chapter of the Quran preceded by a prayer which varies according to the circumstances. See my *Mémoire sur la religion musulmane dans l'Inde,* p. 2.

12. The figure can be found in Mr Herklots' book, plate 1, fig. 2.

13. I mean the tenth day; because the festival is, as a matter of fact, held for ten days, even if they hold it for thirteen days

in the Deccan. The three last days are dedicated to certain additional ceremonies which our author has described.

14. P. 45.

15. See my *Mémoire sur les particularités de la religion musulmane dans l'Inde,* p. 14, 15.

16. See my *Mémoire sur les particularités de la religion musulmane dans l'Inde,* pp. 46 sqq.

17. Karimuddin quoted in the *Travels of Valentia,* vol. 1, p. 477.

18. *Mémoire sur des particularités de la religion musulmane dans l'Inde* p. 52.

19. Ibid., p. 58.

20. Captain; the word *'nakhuda'* is made of *nao,* a ship and *khouaa* [*sic*], God, master, lord.

21. I do not know why Mr Herklots always writes Bharaich.

22. *'Gesu daraz'* is a surname which means long hair. I think there is a reference here to Abd-allah bin Husaini of Kalbargah, author of the commentary, in the *dakhni* dialect of the mystical treatise *Pleasures of Love (Divine)* written by the famous Abdul Kadir Jilani. See Stewart, Catalogue of Tippoo's library, p. 182.

23. Billon, a coin, which is equivalent to the sixty-fourth part of a rupee, that is to say, about 3 centimes.

24. *lota.* A sort of copper pitcher. It is also called *handi,* from which I think comes the expression *inde,* in Provence which has the same meaning.

25. On several of the persons cited here and in the list which is referred to, one can find information in a learned memoir on the lives of *sufis* by Jami, a memoir which M. de Sacy has published in vol. XII, of the *Notices on the Manuscripts of the Bibliothèque du Roi.*

26. *bang,* liquor extracted from the leaves of hemp, which is called in Arabic *warqul khayal,* the leaf of imagination, that is to say, that which excites it and is also called, *hasisul foqara,* the herb of the *faqirs.* See an article on this leaf in the *Chrestomathie arab* by M. de Sacy, 2nd ed., vol. 1, p. 210 sqq.

27. *Observations on the Musulmauns of India; descriptive of their Manners, Customs, Habits, and Religious Opinions, made during a Twelve Years Residence in their Immediate Society,* by Mrs Meer Hassan Ali, 2 vols. in-8, London, 1832.

Appendix

Saints: Real and Apocryphal.

In the revised version of the *Memoir*, de Tassy adds a quotation from *A Thousand and One Nights* to show what Muslims themselves had to say about the spread of their religion—a quotation, which is quite close to the notorious phrase, 'A sword in the right hand and the Quran in the left.'

> When the followers of Jesus abandoned the right path and were so much plunged in heresy and incredulity so as to believe that Jesus was the son of God, the Almighty rejected their religion and created a great Prophet among the Arabs, and gave him a sceptre in the right hand and the Quran in the left, so that people spread over the face of the earth be converted to the only true religion. Inspired by a holy zeal the Prophet named Muhammad, that is to say, 'the Glory', worked powerfully to extirpate polytheism and infidelity. Equally strong in speech and action, he made use of exhortations and miracles. His holy religion is spreading day after day, and we hope that with the passage of time, it shall be the only one reigning in the seven climates of the world, the only true religion which can ensure salvation.[1]

Whatever the intention in introducing this quotation, in de Tassy's *Memoir* it is the saints who rule and kings who are relegated to footnotes. He would have agreed with Schimmel when she says,'The Indian saints contributed more efficiently to the spread of Islam than rulers or official *ulema* '(p.23). The Great Mughal went to Ajmer on foot annually, for several years, out of devotion and not as an obligatory penance as Henry II of England did to the shrine of Becket. Akbar's great grandson Dara Shikoh (1615–59), would make a pilgrimage to Punjab to sit at the feet of the great Hindu saint, Baba Lal Das.

Dara Shikoh and his life and death provide a convenient symbol of the division among Indian Muslims, on the basis of the observance of the religious rule of conduct (*shari'a*) propounded by the Prophet. The basic conflict has been very succinctly presented by Schimmel:

> Dara Shikoh and Aurangzeb in whom the tendencies inherent in Indian Islam seemed to be personified: Dara representing the search for a common basis, a mystical identification between Islam and Hinduism—built upon the first half of the profession of faith, e. g., 'There is no deity save God', to which mystically minded Hindus could as well subscribe, while Aurangzeb, possibly under Naqshbandi influence, laid emphasis on the singularity of Islam as expressed in the second half of the professon of faith: 'Mohammad is the Messenger of God', by which Islam is singled out as the particular religion whose limits are determined by the law brought by the Prophet (pp. 96–7).

It may not be stretching things too far if one applies to the two Princes the terms *be-shar'* and *ba-shar'* usually associated with the saints. Aurangzeb, who got the *Fatawa-yi Alamgiri* compiled, believed that the law brought by the Prophet was complete and final; while Dara Shikoh, who got (through Kayasthas) fifty of the Upanishads translated into Persian, under the title *Sirr-i Akbar* could be accused of looking beyond these limits, hence not committed to the *shari'a*, a *be-shar'*. However, those who do not conduct themselves according to the *shar'* are sinners and not *kafirs*, as Shah Waliullah (1703–62) says:

Everyone who goes to the country of Ajmer or to the tomb of Salar Masud or similar places because of a need which he wants to be fulfilled is a sinner who commits a sin greater than murder or adultery. Is he not like those who call to Lat and Uzza? Only we cannot call them infidel because there is no clear text in the Koran on this particular matter (Schimmel, p.157).

In spite of this injunction and actual prohibition by some Muslim Sultans, devotees, Muslim as well as Hindu, have been visiting the shrines of Muslim saints for about one thousand years; Salar Mas'ud died in 1033. These saints are of all kinds and of various levels of respectibility and are to be found in remotest parts of the country. To what de Tassy says about the saint in the Sunderbans one might add the following about Mubarak Ghazi, whose devotees were ordinary people who never asked themselves whether what they were doing was *ba-shar'* or *be-shar'*: 'Boatsmen and woodcutters, Hindus and Muslims, used to offer him some rice and bananas before entering the jungle where the saint was supposed to ride a tiger' (Schimmel, p. 191).

The most powerful and the most respected among those who conformed more or less to the *shari'a* have been saints who belonged to one Order (*silsila*, chain) or the other. Power and continuity have been retained through the scholarly and spiritual personality of the saint, his anniversary, *urs*, through the formal allegiance to his successor and the feeling of brotherhood among disciples. Over the centuries, these Orders have acquired a distinct character in the matter of rituals, doctrines and also views about contemporary political authority and the non-Muslims. De Tassy refers to three of these Orders, the Chishtiya, the Qadriya and the Suhrawardia. (The Naqshbandis are the fourth most important Order.)

The Chishtiyas are the most respected because there has been a succession of very eminent saints in this Order. The first of any Order to come to India was Khwaja Mo'inuddin Chishti (d. 1236), who was known for his love for the poor and needy. The person, who consolidated and laid the foundation of the main line of this Order was Bakhtiar Kaki (d. 1235), who had met Mo'inuddin in Baghdad but came to India a little later. Kaki's major *khalifa* was Fariduddin Gunjshakar (d. 1257), whose *khalifa* was Nizamuddin Awliya, who made mysticism almost a mass movement. He was succeeded by Chirag-i Dehli (d. 1356) and the latter by Gesudaraz Bandanawaz of Deccan (d. 1422). This Order is distinguished by its saints and the disciples leading an isolated life in *khanqahs*, and not discouraging devotional and congregational music (*sama'.*) They did not insist on the formal conversion of Hindu devotees, and above all they lived on unsolicited gifts and avoided contact with the political authority.

The Suhrawardis, on the contrary, are criticized for being rather too close to government. Baha'uddin Zakaria Multani, who came to India about the same time as Mo'inuddin was, according to Schimmel, 'The richest saint in medieval India' (p. 31). Moreover, succession among them was confined to the family. Baha' uddin's son was not as eminent as his father, though his grandson, Ruknuddin (d. 1335) was. Jalaluddin Husain Makhdum-i Jahanian Jahangasht Suhrawardi (1308–85) was not a *khalifa* but was related to this Order and the Jalalias are supposed to be the followers of his grandfather, Jalaluddin Shaikh Bokhari (Schimmel, p. 33). The Suhrawardis could not be very popular in the country, one of the reasons being their close proximity to political power.

Shah Daula Suhrawardi (d. 1676), de Tassy's Dola, is not a very praiseworthy example of Indian Muslim sainthood. Mrs Hassan Ali's father-in-law appears to have been greatly impressed by the spiritual prowess of this 'Durweish' (vol. II, pp. 300–5) and narrates in detail how this saint was brought in to the presence of the Emperor Shahjahan, who conferred upon him the rare honour of

giving a seat by his side. However, Mujeeb puts him in the
category of saints who were very close to the court.' The
gifts in cash and goods that poured into his *khanqah*
enabled him not only to entertain lavishly but to maintain
a zoo where all kinds of birds and animals, even elephants
and lions were kept' (p. 310). According to Schimmel the
saint had the power to give disobedient parents 'rat-like',
i.e., microcephalic, children, who then may serve at the
shrine and are called his 'rats'(p. 132). However, in a foot-
note, she quotes a writer who suggests that the deforma-
tion is caused by the parents to make the children
successful beggars.

The founder of the Qadria Order, Shaikh Abdulqadir
Gilani (d. 1166), is one of the most respected saints of India,
popular both with the educated and the illiterate. He also
has a month named after him, *'Miran'*, *'Gyarhin'* (the
Eleventh) or *'Bare Pir Saheb'* (The Great or The Greater Pir.)
No vulgar practices are associated with his *urs*, which is
celebrated solemnly in homes with *fatihas* and distribution
of sweets. He died in Baghdad, where he is buried. The
only other great saint who is equally respected is
Mo'inuddin Chisti, but historical accidents could have contributed to his popularity. According to Mujeeb, 'There is
no mention of his (Mo'inuddin's) successors going on
pilgrimage to his grave at Ajmer. His pre-eminence among
the *sufis* dates from Akbar's time' (p. 287). Abdulqadir
Gilani has ninety-nine names, one of them being Miranji.

The name or title Miranji created problems for de Tassy
who made separate entries on this saint (see above, pp.
60–5), confused Abdulqadir Gilani with Shaikh Saddo, and
blamed Roebuck for a mistake the Englishman had not
made.[2] It is regrettable that de Tassy did not correct this
error in the revised version of this *Memoir*, which appeared
as a chapter in *L'Islamisme* in 1874.

Sarwar Sultan or Sultan Sakhi Sarwar did not really
belong to the Chishtiya Order, but Mo'inuddin made him
the Qutub of Multan in interesting circumstances. Accord-

ing to a legend quoted by Qateel (pp. 101–2), Sarwar was
the leader of a group of robbers. One night he tried to gain
entry into the Khwaja's house by breaking the backyard
wall but he did not succeed. Meanwhile, the Khwaja was
informed of the death of the Qutb of Multan and, taking
pity on the unsuccessful robber rewarded him by making
him the Qutb of Multan. But according to the attendants
of the shrine of Baha'uddin Zakaria, another Qutb could
not be buried in the same city where their saint lay and
that is why they said that it was a *chamar's* tomb. This
jealousy was natural, because, as Qateel says, Zakaria's attendants could not have dreamt of the gifts Sarwar's at-
tendant received.[3] Whatever the truth, according to Qateel,
Sarwar was very popular among respectable Hindus and
low-born Muslims. The Hindus might have great respect
for their own saints, but they turned to Sultan Sarwar for
fulfillment of their wishes. There were thousands of *raths*
carrying Hindu devotees; and his *medni* was celebrated
with the same pomp and splendour as that of Madar or
Salar Mas'ud. Hindus attributed all their success in life to
Sarwar Sultan and on Thursdays they distributed *halwa*
specially made for the occasion; in Shajahanabad they
lighted earthen lamps in honour of the saint in one room
of their house (Qateel, pp. 101–2).

Qalandars did not belong to any Order, but were a sect
of *be-shar'* mendicants who disturbed the serious saints and
their disciples. This sect, however, gained respectability
through Shah bu Ali Qalandar (d. 1323), an eminent poet
and mystic. However, in our times, another saint, Lal Shahbaz ('Red-Falcon') Qalandar has become more famous,
thanks to the musical hit, *Dama dam mast Qalandar*.

The special feature of each of the principal mystical orders has very logically been indicated by Shah Waliullah,
who, however, does not mention the Suharwardis:

The *nisba* that I received from Shaikh Abdulqadir Gilani
is purer and subtler; the *nisba* that I received from

Khwaja Naqshband is more overpowering and effective;
the one that I received from Khwaja Muinuddin is
nearer to love, and more conducive to the effect of the
[Divine] Names and the purity of thought (Schimmel,
p. 154).

Salar Mas'ud and Shah Madar

There are many things in common between Salar Mas'ud
(d. 1033) and Shah Madar (d. 1050), who pre-date the saintly Orders and who had no formal successors. Qateel includes both these saints in a chapter entitled, 'Modern
Customs of the Hindus'. As we have already seen the fact
that Salar Mas'ud was a *shahid* and Madar a *ghazi*, did not
affect their popularity among Hindus, who considered
Madar to be Laxman, Ram's brother. Qateel gives us some
idea of the nature of Madar's shrine.

> Since early morning, the attendants of Madar's shrine
> would wait on highways and when they saw a caravan
> approaching at a distance they would run up to it. If
> the travellers were Muslim, they would say, Ali Murtaza,
> Hasan, Husain, Mohammad are all the titles of Madar
> Saheb. And if they are Hindus, they would say, Ram,
> Avatar, Kanhaiyaji and Bhairavn, they are all incarna-
> tions of Shah Madar. Come, visit the shrine. Ask for
> whatever you want and you will receive (pp. 100–1).

However, from what we have read above, it is abundantly clear that it was not devotion alone that attracted
people to shrines. The improper behaviour of women at
Salar's shrine has been noted by de Tassy. 'A remarkably
devout person' told Mrs Hassan Ali that women were dis-
courged from entering the shrine, because those who ven-
tured, 'are immediately siezed with violent pains as their
whole body was immersed in flames of fire' (vol. II, p. 321).
But the prohibition on the entry of women into the shrine
might have been imposed at a fairly late date, because

Badauni had a different experience, one which is worth quoting:

> When Kant and Golah became the *jagir* of Muhammad
> Husain Khan, and, I, in accordance with the decree of
> fate, remained some time in his service, and became *Cadr*
> of that province, and had the responsibility of minister-
> ing to the *faqirs*, on the occasion of a pilgrimage to the
> shrine of that illustrious luminary, that Shaikh of nobles,
> that marvel of truth and religion, Shah Madar (God
> sanctify his glorious tomb) at Makanpur, one of the de-
> pendencies of Qannouj, I, this son of man who have im-
> bibed my mother'crude milk, through the nature of my
> disposition which is compounded of innate carelessness
> (which is the cause of wrong-doing and repentance) and
> of radical ignorance (which conduces to presumption
> and damage, and has descended to me from the father
> of all flesh), wilfully closed the eyes of my intellect, and
> gave it the name of love. So I was captured in the net
> of desire and lust, and the secret contained in the ancient
> writing of fate was revealed, and suddenly in that shrine
> I committed a terrible piece of impropriety. But since the
> chastisement as well as the mercy of God (praise to Him
> and glorious is His Majesty) was upon me, I received
> punishment for that transgression and chastisement for
> that sin in this world, for God, granted to some of the
> relatives of the beloved to overcome me, from whom I
> received nine sword- wounds in succession on my head
> and hand and back.[4]

However, the sainthood of Salar Mas'ud and the very
existence of Shah Madar have been questioned. Badauni
quotes a *shaikh*, who when asked about Salar Mas'ud, said,
'He was an Afghan, who met his death by martyrdom'
(Schimmel, p. 135). About the other saint, Mujeeb says,
'Badi'uddun Madar may be altogether a fiction of the
popular imagination' (p. 287).

Khwaja Khizr

He alone of the apocryphal saints was universally venerated from Peshawar to Chittagong, both by the high-born and the low-born and was not associated with any locality. In Lucknow, according to Mrs Hassan Ali, the festival of the floating of lighted boats (see above p. 134) was called 'Shahbaund' and the boat *Elias ky Kishtee*. Learned people would dismiss it as a *zenanah* festival but Mrs Hassan Ali says that children of all ages participated in it. It was held on each Friday of the last month of the rainy season. The boats would be carried in procession with bands playing music and soldiers marching alongside with it to the river. Her description would corroborate Hodges' above:

> The *kishtee* (boat) is launched amidst a flourish of trumpets and drums, and the shouts of the populace; the small vessel, being first well-lighted, by means of the secreted lamps, on a broad river, in the stillness of the evening, any one—who did not previously know how these little moving bodies of light were produced— might fancy such fairy scenes as are to be met with in the well-told fables of children's books in happy England (vol. I, p. 290).

Mrs Hassan Ali does not mention the name of Khizr but like de Tassy, relates Elias to the biblical Prophets, Elijah and Elisha, with whom she sees a connection.

The fascinating Company painting, 'An Offering to the Ganges,' has been described by Mildred Archer as a 'Hindu procession headed by a band carrying a model boat as an offering to the Ganges'. The group in the picture could be one of the many going to the river-front on a special day for women; men provide only the infrastructure. The women are Muslim, because they are wearing tight pyjamas: Bengali Muslim women would wear saris. Moreover two of the women are wearing *ghaghras*, again a dress worn only by Muslim women from Upper India. They are not from the *'ashraf'*, but from the affluent class of the *'ajlaf'*, who could bear, among other things, the cost of the expensive *bera*. They could have been from the migrant, menial class from Upper India, Calcuttan's term for UP and Bihar. In this connexion it would be worth quoting the whole passage from Afsos, from which a number of words have been left out by de Tassy (see above, pp. 55–6).

> There are a large number of mosques also, but not worth mentioning except the one built by Ramzani Darzi in Suthal Hatti. It is a *pucca* mosque with nine domes, and is the best in the city and definitely beyond the builder's aspiration. There are also a number of Imambaras, because there is not one *sarkar, jamadar, khansaman, nazir* who has not built near his house a two-or-three cubic high plinth with a dome of equal height on it. A few of *chobdars, jamadars* and *bibis* of sahebs have built independent buildings as Imambaras with enclosures. They have spent a lot of money on these constructions, but how can one expect from them taste in construction or a knowledge of the proper ceremonies of *ta'zia-dari*? But if this has been done as an act of merit, they might expect deliverance in the other world, otherwise they are destined to a life of misery and humiliation in both the worlds[5].

As the proper name suggests, Ramzani Darzi (Ramazan-born, tailor) belonged to the lower class, so did the other persons mentioned above. The term *'dupaharia matam'* (p. 56 above) would suggest that these people, or at any rate, the dominant among them were from Bhojpuri region; in Bhojpuri the suffixes, *ya, wa* are added to Hindi nouns. Thus Hindi *do pahar*, becomes *dupaharia*.[6] As for the boat itself, the *ta'zia* and the peacock are almost identical with the drawing given by Herklots (facing page 136) except that in Herklots' sketch, there is a raft instead of two empty *gharas* (pots), and instead of an elephant's head, we have the peacock's tail and the peacock does not have a rosary (*tasbih*) in its beak. Herklots calls it, '*Mohur-punkhee* or *Bayra Kishtee* or *Juhaz.*' '*Mohar punkhi*' was the luxury boat in which rich Bengalis would have an outing. The idea of the *ta'zia* appears to have come from the small rooms on the luxury boat.

Shaikh Saddo

According to Mrs Hassan Ali, this man 'who was very learned but a great hypocrite', is more Faustian than a character from *The Arabian Nights*. The story she had heard in 'the *zeenahnahs* of the Mussulmauns'and recounts in detail, is similar to Roebuck's, which the British lexicographer had taken from Qateel, as the wrong spelling of the Persian '*Ilm-i taksir*'[7] makes it very clear. Even after his death Shaikh Saddo did not leave mortals alone. If a man suffered from some mental illness people would say, 'Ay, it is the spirit of Sheikh Suddoo has possessed him' (vol. I, p. 324). According to Qateel, one could be dispossessed of him only after sacrificing a black goat. Low-class women, elegantly dressed and heavily perfumed, would have formal evening sittings, '*baithak*', when one of them would be possessed by the Shaikh and he would answer through her, questions not only on personal problems but also on public issues. According to Qateel, since even respectable women were possessed by the Shaikh, it could not be said that all such women were frauds. However, he comments, 'In the eyes of the Shi'as, Shaikh Saddo is of very doubtful lineage and even more depraved than Salar Mas'ud, Shah Madar and Sultan Sarwar' (Qateel, p. 106).

Shaikh Saddo is supposed to be one of a group consisting of seven women and seven men; the men are Shaikh Saddo, Zain Khan, Nannhe Mian, Badr Jahan, Chahal Tan, Shah Daryai and Shah Sikandar and the women are Red Fairy, Green Fairy, Black Fairy, Yellow Fairy, Sky Fairy, Sea

Fairy and Light Fairy. They possess women one by one. Sometimes effiminate men from rich families participate in such *baithaks* on Thursday nights and Shah Daryai and Shah Sikandar usually possess them (see Qateel, p. 177).

Goga Pir

It is difficult to give a coherent account of Goga. What follows is a summary of what Qateel has—not very logically—to say about this person in a chapter entitled 'A Description of the beliefs of the Hindus which are outside their *shari'at*.' Khwaja Safa was the master of the art of cleaning, and the patron of the scavengers called Lalbegis. He had framed the rule of conduct of the community and was very close to God. When the Prophet wrote to him inviting him to embrace Islam, he declined and was thus thrown into hell. On his visit to Heaven during *mi'raj*, the Prophet found litter everywhere, and when he was told the reason for it, he pleaded to God to save the Khwaja. The Prophet embraced him when he was released and cleanliness was restored in Heaven.(Qateel points out that the Lalbegis would rather commit suicide than embrace Islam.)[8] Lalbeg, the ancestor of the community had an unusual birth; he fell from the sexual organ of the Khwaja. Lalbegis have a great veneration for Zahir Pir, whom they call also Goga Pir. 'For one whole month, these illiterate people assemble in towns, make a noise with drums and songs, carrying banners and fans made of peacock feathers. Some of them go to Bangar which has the grave of Zahir Pir. It is heard that Zahir Pir was also the son of a Meo and was killed by mistake at the age of eighteen. The Rajputs took pity on him and handed over the body to Muslims (Qateel, pp. 72–4).

In the revised version of his *Memoir*, de Tassy adds the information that Pindari women would worship Goga before their husbands set out to plunder. According to some, Goga was a Hindu, who became Zahir Pir, when he embraced Islam.

Mujeeb (p. 19) quotes from a newspaper (*The Statesman*, 11 March 1959), which carried a story about the temple of Goghaji, a Rajput saint, in Suratgarh, Rajasthan, the worship of whose idol is performed by Muslims. On 17 March, 1992, Door Darshan, the official Indian television network, telecast a documentary on 'Goga meri', in Rajasthan where over eight lakh pilgrims with the usual banners and peacock feather fans visited the saint's shrine. This is managed by Muslims. Among the devotees interviewed were an officer and his wife, who had come all the way from Mizoram, the eastern end of the country.

The Be-shar': *Poets and Faqirs*

The saintly orders had assimilated so much of the extraneous elements that there was hardly anything left for the *be-shar'* to demand respectability, except in poets' circles, where Hafiz and his like were admired and emulated. De Tassy has noted the paradox in the case of Hafiz, who defied the rules of religious conduct yet his devotees thronged to his grave. Mrs Hassan Ali makes a distinction between the two most popular poets in India, Hafiz and Sa'di; the latter was a *'Saalik' sufi* while Hafiz was a *'Majoob' sufi* (vol. II, p. 248). The *majzoob* (literally, soaked, that is, soaked in the wine of divinity) was above the normal rules of moral and social behaviour. As a matter of fact, one of the leading themes of Urdu poetry has been extolling the *be-shar'* life and making fun of the *zahid* (the teetotaller *sufi*), the Shaikh (the *ba-shar'* scholar), and the *muhtasib* (the person who enforced the moral law.) One of the most popular verses on this theme goes:

Don't be misled by my wet clothes; if I squeeze the tail of my shirt, angels would be proud to use the wine for ablution.

There is no doctrinal belief involved in the verse; it is an example of wit, where the figures of speech—Self-praise and Hyperbole—have been used. One can even make a general statement that even though the Urdu poets talk a lot about *sufi* ideas they do not give any evidence of real mystical experience that one finds in the poetry of Hafiz.

Be-shar' life found its popular expression through the travelling *faqirs*, who brought colour in an otherwise dull life. De Tassy mentions four of these sects, Madaris (followers of Madar), Malang (followers of Janamjati, a disciple of Madar), Jalalias (followers of Shaikh Jalal) and Azad, who shaved the whole body including the eyebrows and eyelashes, and were witty and impertinent. The eminent scholar and novelist, Ahmed Ali (1910-94) gives an interesting picture of such *faqirs* in Delhi in the early part of this century:

> They stood before the doors and sang a verse or just shouted for bread or pice or, tinkling their bowls together, they waved their heads in a frenzy, beating time with their feet, singing for all they were worth:
>
> *Dhum Qalandar, God will give,*
> *Dhum Qalandar, God alone,*
> *Milk and sugar, God will give,*
> *Dhum Qalandar, God alone...*

They were ever so many, young ones and old ones, fair ones and dark ones, beggars with white flowing beards and beggars with shaved chins. They wore long and pointed caps, round caps and oval caps or turbans on their heads. And there were beggars in tattered rags and beggars in long robes reaching down to the knees. But they had deep and resonant voices and all looked hale and hearty.[9]

NOTES

1. *L'Islamisme, d'après le Coran l'enseignement doctrinal et la pratique,* Paris, 1874, pp. 290–1.

2. However, Miranji is a title that can create confusion, sometimes deliberately. According to *Firhang-i Asafia* (Delhi, 1987, vol.II, pp. 2259–60), Miranji is the title of Shaikh Saddo as well as that of Khwaja Mo'inuddin Chisti and Abdulqadir Gilani. While *Miranji ka chand* (the moon of Miranji, that is, Rabi'ul akhir) is dedicated to Abdulqadir Gilani, *Miran ka bakra* (Miran's goat) and *Miran ki karhai* (Miran's cauldron) refer to the edible meant for the *fatiha* of Shaikh Saddo. Khing Sawar, Mo'inuddin's father-in-law was also called Miranji. According to P. M. Currie, in Ajmer there are two shrines dedicated to Abdulqadir Gilani and Husain Khing Sawar respectively and the notorious Shaikh Saddo is also believed to be buried there. 'It, therefore, seems reasonable to speculate that this association of Miran Saheb ... (Khing Sawar), with the Miran Sahib of the folklore (Shaikh Saddo) of northern India and even more remotely with 'Abd al-Qadir Jilani, enabled the cult to establish itself and provide a useful additional income to the inhabitants of Taragarh Hill (in Ajmer where Mo'inuddin Chisti is buried.) (*The Shrine and Cult of Mu'in al-din Chishti of Ajmer*, 1989, p. 121).

3. About the jealousy of *pirs* regarding their respective geographical area there is an interesting anecdote. Fariduddin went to Multan, an acknowledged centre of learning and became the disciple of Qutbuddin. But the city was also the seat of Baha'uddin Zakariya.'Once when master and disciple met Shaikh Baha'uddin in the mosque, the latter set right their shoes as they were leaving. This was a hint that they should go elsewhere and leave Multan to Shaikh Baha'uddin' (Mujeeb, p. 137n).

4. *Muntakhabu-t-Tawarikh* by Abdul Qadir Ibn-i-Muluk Shah known as Al Badaoni, pp. 140–1, vol. II, trans. and ed. by W.H. Lowes (Patna, 1973).

5. *Araish-i Mahfil* (Delhi, 1945, p. 146).

6. Compare the refrain of an old Bhojpuri song, '*suna suna ho chameli, kahai ghumelu akeli dupaharia main.*' 'Listen, O jasmine-bodied one, why do you wander lonely in the midday (sun)'.

7. The 'scribe's devil' has created a great deal of confusion about *Ilm-i taksir*. In the Persian original of *Haft Tamasha*, the word given appears to have been spelt with a *se* instead of a *sin*, which Roebuck accepted; hence de Tassy, not finding the word in a dictionary, guessed that it could have been *ilm-i taqdeer*. *Ilm-i taksir* is synonymous with *ilm-i jafr*, which has been defined in the following words in Steingass' *Persian–English Dictionary*: 'The art of divining from certain characters written by (Hazrat) Ali upon a camel's skin which contains all events, past, present and future.'

Thus Roebuck is correct, but only partly, when he says that *ilm-i taqsir* is the art of foretelling the future. De Tassy's guess was a wild one, because there is no such thing as *ilm-i taqdeer*, art of fate.

8. There is another legend about the origin of Lalbegis, which would indicate how the 'low-born' managed to retain their identity in the hostile world of the 'high-born', whether Hindu or Muslim. One day when Balmikji, the old sweeper in *Mi'raj* was returning to his home in Ghazni, God gave him a folded shirt, in which the baby Lal Beg was discovered. God called him Noori Shahbala, who, along with his community would not eat pork because his foster-mother was a sow. He made a separate masjid of a brick-and-a-half, *dairh int ki masjid*, and according to the author of *Farhang-e Asafia*, who has recorded this legend, this practice was still followed at the time of writing of the *Firhang*, that is the beginning of the twentieth century (*Farhang-e Asafia*, vol. III, p. 1920).

9. *Twilight in Delhi*, New Delhi, 1973, p. 17.

Select Bibliography

Works by Garcin de Tassy

Les oiseaux et les fleurs, allégories morales d'Azz-ed-din mocadessi, publiées en arabe: avec une traduction et des notes (Paris, 1821).

Coup d'oeil sur la littérature orientale; discours lu au cercle des Arts, (Paris, 1822).

'Exposition de la fois musulmane, traduite du turk de Mohammed ben-Pir-Ali-Alberki, avec des notes, etc.', *Journal asiatique*, 1823.

'La Caravane, séance de Hariri, traduite de l'arabe', *Journal asiatique*, 1824.

'Conseils aux mauvais poètes, poème de mir Taky, traduite de l'hindoustani', *Journal asiatique*, 1825.

Doctrine et devoirs de la religion musulmane, tirés textuellement du Coran, suivis de l'eucologe musulman traduit de l'arabe, (Paris, 1826).

'Notice sur les fêtes populaires des Hinduous d'après les ouvrages hindoustani', *Journal asiatique*, 1834.

Les aventures de Kamrup (Paris, 1834).

Oeuvres de Wali, publiées en hindoustani (Paris, 1834).

'Mazih-i Curan, c'est-à-dire, l'Exposition du coran (le Coran en arabe, accompagné d'une traduction interlineaire et de notes marginales, en hindoustani', Calcutta, 1829, published in *Journal des savants*, July 1834.

'Notice sur des vêtements avec des inscriptions arabes, persanes et hindoustani', *Journal asiatique*, 1838.

'Analyse d'un monologue dramatique indien', *Journal asiatique*, 1850.

Mémoire sur les noms propres et les titres musulmans (Paris, 1854).

Chants populaires de l'Inde, traduit par M. Garcin de Tassy (Paris, 1854).

Discours prononcé à la séance publique annuelle de la société d'ethnographie (2nd ed., Paris, 1867).

Histoire de la littérature Hindouie et Hindoustani (New York, 1870, reprint 1968), 3 vols.

Maqulat-i Garcin de Tassy, tr. Dr Hamidullah (Karachi, 1975).

La langue et la littérature hindoustani de 1850 à 1877 (Paris, 1878).

Secondary Sources

Afsos, Shair Ali, *Araish-i Mahfil* (reprint Delhi, 1945).

A. P. Badaoni, Abdul Qadir Ibn-i Muluk Shah, *Muntakhabat-i Twarikh* (reprint Patna, 1973).

Ahmad Ali, *Twilight in Delhi* (reprint New Delhi, 1973).

Arberry, A. J., *Classical Persian Literature* (London).

Balbir, Nicole, 'De fort William au Hindi Littéraire: la transformation de la khari Boli en Langue Littéraire au XIXe siècle', in *Littératures medievales de l'inde du Norde*, ed. Françoise Mallison (Paris, 1991).

Crooke, William, *Qanoon-i Islam* (reprint, New Delhi, 1972).

Currie, P. M., *The Shrine and Cult of Mu'in al-din Chishti of Ajmer* (New Delhi, 1989).

Dubois, J. A., *Moeurs, Institutions et ceremonies des peuples de l'Inde* (Paris, 1825).

Khudabukhsh, Salahuddin, *Essays: Indian and Islamic* (London, 1912).

Masud Hassan Rizvi Adeeb, *Fa'iz Dehlvi aur Diwan-i Fa'iz* (Aligarh, 1965).

M. Mujeeb, *Indian Muslims* (New Delhi, reprint 1985).

Mohammad Umar, *Haft Tamasha* (Delhi, 1968).

Nott, C.S., *The Conference of the Birds: Mauliq Ut-tair; a philosophical and religious poem in prose/Farid Ud-Din Attar: rendered into English from the literal and complete translation of Garcin de Tassy* (London, 1961).

Schimmel, Anne-Marie, *Islam in the Indian Subcontinent* (Leiden, 1980).

———, *Islam in India and Pakistan*, Leiden, 1982.

Vaudeville, Charlotte, 'BARAHMASA: les chansons des douze mois dans les littératures indo-aryennes' (Pondicherry, 1965).

Index

The words in brackets with an asterisk are de Tassy's version of non-European terms.